NONPOINT SOURCE POLLUTION REGULATION: ISSUES AND ANALYSIS

FONDAZIONE ENI ENRICO MATTEI (FEEM) SERIES ON ECONOMICS, ENERGY AND ENVIRONMENT

This series serves as an outlet for the main results of FEEM's research programmes in the areas of economics, energy and environment.

Nonpoint Source Pollution Regulation: Issues and Analysis

edited by

Cesare Dosi
University of Padova,
Department of Economics,
Padova, Italy

and

Theodore Tomasi
Michigan State University,
Department of Agricultural Economics,
East Lansing, MI, USA

KLUWER ACADEMIC PUBLISHERS
DORDRECHT / BOSTON / LONDON

A C.I.P. Catalogue record for this book is available from the Library of Congress.

ISBN 0-7923-3121-4

Published by Kluwer Academic Publishers,
P.O. Box 17, 3300 AA Dordrecht, The Netherlands.

Kluwer Academic Publishers incorporates
the publishing programmes of
D. Reidel, Martinus Nijhoff, Dr W. Junk and MTP Press.

Sold and distributed in the U.S.A. and Canada
by Kluwer Academic Publishers,
101 Philip Drive, Norwell, MA 02061, U.S.A.

In all other countries, sold and distributed
by Kluwer Academic Publishers Group,
P.O. Box 322, 3300 AH Dordrecht, The Netherlands.

Printed on acid-free paper

Printed in the Netherlands

Table of Contents

Foreword

In April 1992 the Foundation Eni Enrico Mattei organized a workshop on the regulation of nonpoint source pollution. This volume includes the proceedings of that meeting, as well as additional original contributions, in an attempt to provide an overview of recent theoretical developments in the field.

Research on the causes, consequences, and control of nonpoint source pollution has been carried out over the last two decades. Interest in this subject has grown as a result of the increasing recognition of the insufficiency of traditional pollution control policies focused on the large scale, confined, and generally predictable pollutant discharges. In fact, many contemporary problems are caused by the combined activities of small polluters, along with natural processes, intermittent and unpredictable events, and often involve pollutants with complex environmental outcomes.

Despite the progress made in understanding the nature and size of pollution from diffuse sources, the issue of regulation is still far from being systematically and adequately addressed. This policy vacuum is partly attributable to the difficulty of adapting the traditional point source regulatory tool kit to the specific features of nonpoint source problems. Such features include the technical difficulty of identifying sources and measuring individual emissions, their variability over time and space, the role played by natural processes in determining pollutant discharges at source and their ultimate impacts on the receiving environmental media.

A common theme in the essays included in this volume is the attempt to elaborate regulatory solutions starting from the explicit recognition of the problems of incomplete information encountered when dealing with nonpoint source pollution regulation. Such problems, which can be directly or indirectly traced back to the difficulty of identifying sources and measuring individual discharges, imply that, rather than monitoring emissions at source, alternative solutions have to be found as a basis for a regulatory policy. In this sense many of the considerations made in this volume not only refer to problems generally classified in technical literature as nonpoint source pollution, but are also applicable to many phenomena labelled as point source problems in the presence of weak or inadequate monitoring capability. This applies, in particular, to those situations in which, despite the fact that the technical means exist for measuring the volume and quality of emissions, widespread and continuous monitoring of individual discharges is economically or administratively unfeasible, and alternative ways of establishing cause and effect links between economic activities and observed environmental damages are required.

Domenico Siniscalco,
Director Fondazione Eni Enrico Mattei

Preface

The belief that adequate pollution control can be achieved by managing sources that are easy to identify, and discharges that are susceptible to "end-of-pipe treatment", has been undermined by on-going evidence of significant pollution phenomena that are due to nonpoint sources, largely untouched by regulation.

In the technical literature and official documents the term nonpoint source (NPS) pollution usually refers to land-based and airborne, diffuse sources of pollutants affecting aquifers, lakes, rivers and coastal ecosystems. However, sources affecting other environmental media, such as air pollution caused by automobiles, share many of the features characteristic of NPS pollution. Consequently, its scale and relevance goes well beyond what documents that focus on water quality problems would lead one to believe.

The available data regarding water quality are sufficient, however, to demonstrate the significance of NPS pollution. Although this information is patchy, and the sources, types and amounts of pollutants vary among countries and geographic regions involved, it is possible to draw some general conclusions.

NPS's have been identified as the major contributors to sediments, nutrients, pathogenic bacteria, and pesticides. Agriculture is generally considered the most pervasive cause of NPS water quality problems. Pollutant loadings from urban runoff are next in the order of importance. Additional sources include construction sites, mining areas, forestry operations, highways, and waste disposal sites.

According to the 1986 National Water Quality Inventory prepared by the U.S. Environmental Protection Agency, NPS's are the main causes of the remaining surface water quality problems in the United States. In all but seventeen states, NPS problems exceed any other category of pollution in impaired rivers and streams, cause the predominant pollution in seventy-six percent of lake acres not meeting water quality standards, and are responsible for the primary pollution effects in nine of the sixteen states reporting impaired estuarine waters.

Underground water resources, which provide a large proportion of drinking water in many countries, are also increasingly at risk. According to a report prepared in 1987 by OECD, in Member countries, there are no large-scale sources of specific pollutants – such as nitrates and pesticides – other than diffuse sources.

Studies conducted within single European countries have reached similar conclusions. For example, a recent investigation carried out in Italy, aimed at assessing the relative role played by various sources of the serious eutrophication problems in the Venice Lagoon, estimated that approximately 75% of the total amount of nutrients released within the watershed can be attributed to agricultural and urban diffuse sources.

Despite the recognition of their scale and pervasiveness, NPS pollution problems are still far from being comprehensively and adequately addressed in industrialized countries, let alone in developing ones. In the United States, the first country to address officially NPS pollution problems, only in 1987 did Congress announce its willingness to shift from fifteen years of NPS problem identification to an action program. Even this program, however, is still far from being fully implemented. In the European Community, to our knowledge, the first official document which explicitly mentions NPS pollution is a directive proposal prepared by the EC Commission in 1989, addressing water quality problems caused by nitrate releases from crop land and urban areas.

In many countries, pollution control policies, and water quality regulation in particular, still generally require direct monitoring of pollutant discharges at their source. The Italian case is emblematic in this respect. According to current Italian legislation on water pollution control, mandatory technological standards and sanctions for non-compliance apply only if the pollutant sources are localized and discharges may be measured upstream of the pollutant receptor point(s). The result is that, even though economic activities involve pollution of water bodies, they are exempted from regulation if pollutant loadings are not technically detectable and measurable so as to establish that legal emission standards are or are not met.

There are various explanations for the absence or modest impact of NPS pollution control actions.

One explanation surely lies in the fact that only recently has the relevance and widespread nature of the phenomenon been recognized. NPS pollution was first considered to be a problem only about twenty years ago.

Second, some of the economic activities that are responsible for NPS pollution – agriculture in particular – were for many years more or less deliberately exonerated from environmental regulation.

A third explanation lies in the difficulties faced by policy-makers in modifying a regulatory approach directed to control point sources. Pollution abatement traditionally has relied on enforcement, compliance with strict effluent standards, and specific mandatory pollutant treatment equipment. Extending this approach *tout court* to address NPS pollution may be impossible due to the defining characteristics of NPS pollution problems.

Despite the lack of incisive and widespread control and regulation, research on the causes, consequences and control of NPS pollution has been carried on over the last two decades. These studies have led to identification of the underlying features of NPS problems which distinguish them from traditional point source ones.

One significant aspect of NPS problems is the need to prevent the formation of pollutant loads at their source. It is difficult to rely on structural devices controlling subsequent pollutant delivery. This is not to say that "add-on" technological devices to intercept and neutralize pollutants running-off specific NPS's are not available in general. For example, concerning effluents from cultivated land, some structural correctives have been proposed, such as setting up treatment areas where the impact of some pollutants is softened before delivery. With regard to urban runoff, end-of-pipe treatment devices (such as detention storage facilities combined with physical or biological pollutant treatments) have been developed; site control measures, such as urban drainage (re)planning are also available, and, depending on the specific situation at hand, may contribute to reducing harmful environmental impacts associated with urban activities.

However, at the present state of knowledge structural correctives are not generally regarded as suitable options, and even when technically feasible, they may be prohibitively expensive to implement. It should also be noted that some of the correctives proposed – such as the replanning of urban drainage systems combined with end-of-pipe treatment – would transform the area as a whole into a point source; nevertheless, pollutant activities operating within the area would retain their nonpoint nature for the local authority responsible for designing and managing pollutant treatment facilities. Moreover, the structural correctives developed to deal with specific NPS problems tend to focus on pollutant delivery control relative to selected receptor points, but do not generally address the harmful impacts of NPS loadings which may occur during the movement of pollutants from their sources to final receptor points.

A second important feature of NPS pollution is the part played by the physical characteristics of the site where the polluting activity operates, as well as the area through which pollutants move, in determining emissions at source and their ultimate impacts at receptor points. For example, economic activities sharing the same production processes may give rise to significantly different levels of emissions and environmental impact depending on their on-site physical endowments and their location in relation to affected environmental media.

This aspect is not an exclusive feature of NPS problems. Even for some point sources the polluters' "typology" (for example the relative efficiency of machinery and equipment) and location may affect the extent of emissions and their ultimate environmental impacts. However, there are two features which characterize NPS problems: i) the sheer number and variety of sources, and ii) the likelihood that on-site characteristics are significantly affected by exogenous and unforeseeable events, such as weather conditions.

The third and probably most nearly defining feature of NPS problems is the difficulty of identifying sources and measuring individual emissions: while point sources enter the environment at discreet, identifiable locations, NPS pollution does not result from releases at a specific, single location (such as a single pipe) and pollutants are usually dispersed by natural processes. Inferring individual responsibilities from observation of pollutant concentrations at selected receptor points is also difficult: while pollutants from point sources are usually delivered more or less proportionally to releases at the source, NPS pollutants may follow tortuous transport paths before delivery, and may qualitatively change during transport.

The difficulty of employing on-site pollutant treatment facilities makes preventive emission control measures the preferable, and sometimes the only viable option for dealing with NPS problems. From a technical point of view, the *ex ante* abatement measures required have been identified and, although they vary according to the specific problem at hand, the proposed correctives typically call for changes in production or consumption processes.

Obviously, the required changes in land uses and management practices are unlikely to be implemented autonomously and need to be stimulated through appropriate regulatory instruments. Regulation – generally required for all significant pollutant activities – appears to be all the more necessary in NPS cases. This is primarily due to the fact that the impact of the discharges is often seen in environmental media that are distant from the sources, so that the polluters and the beneficiaries of pollution abatement are geographically separated. In such circumstances, polluters are less likely to adopt the required abatement actions spontaneously.

Economists dealing with spatial environmental externalities have developed a large body of literature on this subject, providing a broad menu of policy instruments. These range from technological and performance standards to more market oriented mechanisms. It is well known, however, that, especially in the past, the instruments proposed and evaluations of their relative efficiency and feasibility have been based on fairly stylized situations. A recurring, often implicit, assumption is that individual emissions may be readily inferred or observed at zero or reasonable costs. In the case of NPS pollution, this assumption is unjustified, and should be abandoned in order to identify and evaluate NPS control policy instruments.

More generally, NPS control cannot be properly addressed without explicitly accounting for the information problems likely to be encountered when designing and implementing regulatory instruments. This is not to say that in the past, the environmental economics literature has not taken problems of incomplete and asymmetric information into account. In fact, the economists' preference for "price mechanisms" (effluent taxes or marketable emission permits) over command-and-control measures, is based on the assumption that polluters possess better information about abatement costs. Regarding NPS polluters, however, the information problems encountered do not necessarily have to do with the environmental authority lacking knowledge about abatement costs, at least as traditionally stated. Rather, informational aspects of the regulatory problem for the NPS's vary markedly across specific situations and policy approaches adopted.

Most of the authors dealing with NPS pollution base their analyses on the difficulties of either directly observing emissions at their source, or inferring individual responsabilities from observable ambient pollutant concentrations. These authors then focus on developing alternative regulatory instruments not based on an assumption of monitorable individual discharges.

Some contributors consider that one alternative is to be found in the total amount of pollutants that actually have been delivered and which are generally observable at reasonable cost at selected receptor points. Policy instruments consistent with such a "direct regulatory approach", mainly take the form of "nonpoint" tax/subsidy schemes that depend on deviations between measured and desired ambient pollutant concentrations. The rationale behind this proposal is that by setting an incentive mechanism based on an observable variable (ambient pollutant concentrations) the regulator would induce certain unobservable actions (individual emission at source abatement).

Other authors, in contrast, employ estimated, as opposed to observed, emissions. Such estimates are obtained by means of deterministic or probabilistic bio-physical models of NPS pollution. Policy instruments consistent with such an "indirect regulatory approach" range from management practice standards to input/output taxes. The rationale behind these proposals is that, by granting regulatory legitimacy to such predictive models, the regulator would artificially transform, *ex lege,* NPS activities into point-source ones.

Both these general strategies of intervention are taken up in various ways in the contributions proposed in this volume. Each one assumes the existence of different information structures, and would thus evoke specific additional information problems when designing policy instruments consistent with the overall strategy.

The authors who have elaborated and analyzed policy instruments consistent with the "direct" regulatory approach, tend to assume that polluters possess superior information on the nature and extent of emissions at their source, as well as the relative efficiency of abatement management practices. Authors who investigate the second strategy of intervention tend to assume that polluters do not hold such a private information, and agents are unable to make correct evaluations of their ultimate adverse environmental impacts due to the complexity of pollutant transport processes.

As mentioned above, the direct and indirect regulatory approaches may lead to further informational problems, again ultimately tied to the difficulty of directly monitoring individual emissions. For example, the adoption of the "indirect" approach requires observation of the polluters' management practices and other relevant variables such as the physical features of their site of operation. In many cases such data are not collected at all, or when collected they are different from those needed to feed the available NPS mathematical models. In such situations, adopting the indirect regulatory approach will imply moral hazard and/or adverse selection problems.

Besides specific policy recommendations, we feel that the most important positive externality produced by the economic literature on NPS pollution control lies in explicit attempts to address information problems which have been somewhat ignored or understudied in past research in environmental economics. In this regard, it is worthwhile drawing attention to the fact that many information problems pointed out by authors dealing with NPS pollution are also encountered to varying extents when addressing "traditional" point-source problems. This conviction not only stems from the fact that pollutants generated as point sources are often quickly dispersed in the environment, hence acquiring some

NPS features, but also from the fact that, even when technical means exist for measuring the volume and quality of individual emissions, widespread and continuous monitoring may not be economically viable. Indeed, the literature on optimal monitoring and enforcement might be viewed as a bridge between the pure point source problem and the NPS problem.

The increasing interest shown by economists in NPS regulation is testified by the number of recent empirical and theoretical papers on this subject. A sample of these contributions has recently been collected in a book edited by Clifford Russell and Jason Shogren for Kluwer Academic Publishers. Relative to this previous volume, the current volume places somewhat greater emphasis on theoretical issues related to NPS regulation. This reflects an attempt – shared by most of the current contributions – to link NPS policy design to recent developments in economic theory, such as the economics of uncertainty and information.

The first chapter, written by Ted Tomasi, Kathleen Segerson and John Braden, provides a survey of the problems encountered in NPS pollution control from an economic perspective. The authors propose a model aimed at integrating NPS pollution features and various contributions from the literature within a common framework. Some of the issues raised and discussed in this chapter are then developed in the other contributions included in the volume.

In the second chapter, Jean Jacques Laffont proposes a general framework which provides a conceptual and formal link between NPS policy design and the literature on regulation under incomplete information. The author takes as its starting point the regulation of a natural monopoly which creates pollution that can be abated through a costly technology that modifies the relationship between the level of production and the level of pollution. After examining the case where individual emissions are observable, Laffont turns to a situation where the regulator is unable to monitor the pollution generated by different plants within the same firm or different firms, and identifies conditions when the unobservability of individual emissions does not create social costs. The final part of the chapter is devoted to the optimization of pollution in an industry, when, besides individual appropriate levels of pollution, regulation must also account for the number of firms and/or their location.

In the third chapter, Anastasio Xepapadeas explicitly adopts a "direct" NPS regulatory approach. A problem involved in adopting this approach is that fiscal instruments consistent with it, such as "nonpoint taxes" which depend on deviations between observed and desired total ambient pollution concentrations, would apply to every agent, irrespective of his or her individual contribution. This might undermine their political viability. Assuming that the evaluation of individual emissions may be improved by investing in monitoring facilities, Xepapadeas then analyzes optimal policy rules, comprised of investment in monitoring equipment, and taxes on individual emissions coupled with nonpoint penalties.

The fourth contribution, by Ramu Govindasamy, Joseph Herriges, and Jason Shogren, also takes as its starting point the concern with avoiding the possibility that agents adopting abatement management practices should have the same probability as non-complying agents of being penalized whenever observed concentrations exceed ambient pollution standards. These authors develop a regulatory approach based on rank order tournaments (ROT) which may contribute to overcoming this problem. After describing how ROT can be used as an environmental policy tool, Govindasamy et al. discuss the implications of and possible caveats to the use of ROT in NPS pollution control.

The fifth chapter, prepared by Cesare Dosi and Michele Moretto, employs the "indirect" NPS regulatory approach, in that they assume individual contributions to ambient pollutant concentrations are evaluated according to a deterministic bio-physical model. Among the polluters' characteristics affecting the extent of (estimated) discharges, the authors focus on the physical features of the firms' site of operation ("site quality"). Such features are assumed to evolve randomly over time, with uncertainty growing along with the time horizon. However, polluters have access to a technology which allows site quality to be "pushed upward" through lump-sum maintenance expenditures which only apply when site quality hits a predetermined lower barrier. The chapter then concentrates on the environmental authority's optimal "barrier policy", comprising minimum site quality standards and subsidies for the maintenance activities undertaken by polluters.

Similarly to the previous contribution, in the sixth chapter James Shortle and David Abler consider the case where, although individual emissions are unobservable, a probabilistic NPS model is available, allowing the regulator to formulate expectations conditional upon observation of the polluters' management practices and other relevant variables. After examining the case where the authority possesses perfect information about the polluters' relevant specific characteristics, the authors turn to a situation where these characteristics are private knowledge, and review alternative incentive mechanisms, including polluters' specific tax schemes, and mixed instruments consisting of taxes, subsidies and permits for the use of potentially polluting inputs.

The final chapter, prepared by Ted Tomasi and Arthur Weise, focuses on some spatial concerns relevant to pollution control. Unlike many contributions in this area, it considers issues of competition between land uses in a long-run equilibrium setting. Of particular concern is the ability of pollution control instruments to account adequately for the land use changes they induce, an ability which is seen to depend on complex aspects of the transportation of pollutants through space. Although the model is not a nonpoint model (standard Pigouvian taxes on emissions are considered), the authors do raise many questions that need to be addressed in future research in the NPS setting, and draw some conclusions in this regard.

Chapters 1, 2, 3, 4, and 7 consist of revised versions of papers presented during the workshop on "NPS Pollution: Issues and Analysis", held in April 1992 in Milan at the Fondazione ENI Enrico Mattei, which provided financial support and sponsorship. Original contributions prepared by Shortle and Abler, and by Govindasamy et al. have been included in an attempt to make the work more complete.

We wish to thank Domenico Siniscalco and Carlo Carraro, respectively director of the Fondazione ENI and director of the Foundation's research project "Energy and environment: markets and policies", whose encouragement was essential for preparing and publishing this volume. A special acknowledgement is offered to those who attended the workshop and referees who provided useful comments and criticisms. We end with expressions of sincere gratitude to the Foundation's Staff, and in particular Giorgio Barba Navaretti, Alessandro Lanza, Rita Murelli, and Roberta Ranzini, who assisted us with competence and kindness.

Cesare Dosi and Theodore Tomasi

1. Issues in the Design of Incentive Schemes for Nonpoint Source Pollution Control

THEODORE TOMASI*, KATHLEEN SEGERSON** AND JOHN BRADEN***

**Michigan State University; **University of Connecticut; ***University of Illinois*

1. INTRODUCTION

A large body of economic literature has addressed the question of how to design policies to control environmental pollution efficiently. Until recently, the vast majority of this literature was concerned with point sources of pollution where the regulator can observe the pollution emissions of individual agents (perhaps at some cost) and there are no important stochastic influences on the system. Under these conditions, it is well-known that a system of Pigouvian taxes, quantity standards, or transferable discharge permits can in principle be employed to internalize the pollution externality and sustain a Pareto efficient allocation in the economy. For a thorough analysis in this spirit, see Spulber (1985).

While considerable progress has been made toward reducing point source pollution problems, a substantial proportion of the remaining pollution difficulties are due to nonpoint source (NPS) pollution. In the NPS setting, the assumption that the regulator can observe individual emissions at reasonable cost is untenable. For this reason, the standard policy prescriptions used for point source pollution (emission taxes, quantity constraints, or transferable permits) cannot be used to control NPS pollutants.

In this paper we will outline the key regulatory difficulties that arise in the control of NPS pollution. As will be seen, these issues derive from the existence of uncertainty in the overall social and biophysical systems of interest and from the important role of asymmetric information among regulators and polluters.

C. Dosi & T. Tomasi (eds.), Nonpoint Source Pollution Regulation: Issues and Analysis, 1–37.

While we hope to provide a framework applicable to many NPS problems, for convenience we will use the effect of agricultural pollutants on a water body as an example throughout.

The main goal of this paper is to assess our understanding of the role of uncertainty and asymmetric information in NPS pollution control. We first present an overview of the many issues that arise when modelling the NPS regulatory problem. We then present a stylized model that captures many of these important features. Even in a highly stylized model, the considerations we set forth result in a very complex situation. We offer the model not for analysis of all relevant issues but rather as a means of integrating these issues within a common framework. We hope to illustrate how the various pieces of the extant literature fit together, as well as identify gaps in the literature as we see it.

2. THE NONPOINT SOURCE POLLUTION PROBLEM

2.1. *The Basic Setting*

By definition, the sources of NPS pollution (either within a given firm or across firms) are diffuse rather than concentrated, and combine to generate environmental damages that are borne by others. In some cases, at least part of the effects may be borne by the polluters themselves and thus internalized. For example, farmers may drink from wells that are contaminated by their own applications of pesticides or fertilizers. However, in most cases, most of the impacts of NPS pollution will not be internalized. Also, in some cases the external effect may be positive. An example is when farm chemicals "contaminate" groundwater which is used for irrigation; a neighboring farmer benefits by irrigating with chemical-enriched water. These positive effects are relatively rare, and we will not consider them further here.

Important differences can arise in modeling the NPS phenomenon regarding the number of firms contributing to ambient pollution at a given location, the potential contribution of each firm (as determined by location differences), technological heterogeneity across firms, and the role of random variables in determining ambient pollution and production. Variations in these features of the problem give rise to different regulation problems.

2.1.1. Number of Firms

In some of the NPS literature it is supposed that there is only one polluting firm, (e.g., Dosi and Moretto, 1990, 1991) in a setting with adverse selection and no moral hazard. In this case, the regulator is designing a contract for a single firm of unknown type, with interactions among firms not taken into consideration. This is appropriate, for example, when there is only one farm located near a waterway where agricultural chemicals are causing concern, or where many firms are sufficiently independent of one another to allow them to be regulated individually. The simplifications that result from having only one potential polluter are substantial. Even if that polluter has numerous diffuse sources (such as several fields), there is only one private decision-maker whose decisions determine emissions. In addition, there is no question of "responsibility" for observed ambient pollutants, although information problems relating to firm type may still exist.

Unfortunately, most NPS problems involve multiple polluters[1] whose (diffuse) emissions combine to determine a single measure of ambient pollution at a given location. Moral hazard models and adverse selection with multiple polluters are discussed in, respectively, Segerson (1988) and Xepapadeas (1991, 1992) and Shortle and Dunn (1986). The existence of multiple polluters raises a number of difficult regulatory issues, most of which relate to information. For example, it is no longer possible to attribute ambient pollution to the activities of any one firm since "damages" are not separable across firms. Thus, it is necessary to have information about each firm's potential contribution. The larger the number of firms, the more difficult is the information problem for both the regulator (obtaining information about firms) and the firms themselves (obtaining information about each other). In addition, the existence of more firms implies a greater potential free-rider problem, if each firm perceives its own pollution contribution to be small relative to the group. Finally, an increase in the number of firms decreases the likelihood of cooperation among firms to reduce ambient pollution levels. For example, if all individuals who use a local water body are also farmers whose activities contribute to the deterioration of its water quality, there may be a greater possibility of cooperative behavior to reduce contamination since the parties who bear the cost will also reap the benefits.

2.1.2. Space

Incorporating spatial heterogeneity among firms is crucial in the NPS setting. Spatial heterogeneity is, of course, important in other contexts as well. For example, there are the familiar influences of location differentials on the costs of transportation and of land quality on rents. While these issues are relevant here as well, they are not essential to the NPS pollution problem. The characteristics of many pollutants are altered as they move from their point of origin (a "farm field"). Hence, the impacts of a firm's emissions on the environment are altered with the location of the firm relative to the point where environmental impacts are determined. As a result, two identical activities conducted at different locations will have different environmental impacts. For example, pesticide applications on farms that are located close to water bodies are likely to have a greater impact on water quality than are applications on more distant farms.

These location differences imply different costs of changing environmental harm at some other point in space. Efficient environmental regulations recognize these cost differentials. Production (and perhaps also prices) needs to be indexed by location. There is an important distinction between the indices: "at the edge of the farm field," "at the stream," and "at the point(s) where pollution damages are realized."

Several papers have been directed to the implications of second best pollution policies in a spatial setting (e.g., Park and Shabman, 1982, Braden *et al.*, 1989, Kozloff *et al.*, 1990). These papers recognized that agricultural pollution is spatial in nature but that extant policies, such as best management practices or fixed emission standards for erosion, are applied uniformly. Research has been conducted to assess the benefits of increasing the spatial heterogeneity of instruments ("targeting") so as to better equate across emitters the marginal cost of improving ambient quality at a monitoring site. The results suggest substantial benefits from using spatially-differentiated policies.

In the theoretical literature on NPS incentive mechanisms, spatial heterogeneity is typically incorporated simply by letting the effect of abatement on ambient pollution vary across firms. There has been little explicit modeling of the spatial dimension of the problem (Laffont 1992).

2.1.3. Technologies

Farms differ considerably in their technologies, where "technologies" here refers to both crop production technologies and pollution technologies (i.e. the impact of their input or output levels on ambient pollution at a given location). These differences stem from a number of factors, including climate, soil type, slope, depth of groundwater and intervening geologic structures, etc.. For example, farms with high quality (fertile) soils are likely to have lower production costs than farms with lower quality soils. Likewise, farms in steeply sloped areas with extensive runoff will have greater potential impacts on surface water quality than farms on flat land where there is less runoff. In many NPS models, it is assumed that there is some (convenient for the modeler) correlation between productivity and pollution features on all farms. In addition, there may be several environmental implications of one feature. For example, when runoff is reduced, leaching may be increased; thus, farms on flat land may contribute more to groundwater contamination than farms in steeply sloped areas. This potential tradeoff between surface water pollution (from runoff) and groundwater pollution (from leaching) has been widely-recognized, but not as yet explicitly been embodied in policy.

The technology defines a firm's "type." Differences in firms' production technologies are reflected in cost or profit functions that vary across firms, while differences in pollution technologies imply firm-specific abatement cost functions. Note that by abatement, we mean changes in the flow of pollutants from the firm; differences is abatement cost may be augmented by spatial heterogeneity, which affects the impact of abatement efforts on ambient environmental quality.

One implication of technological heterogeneity and the large number of firms is the inability of the regulator to know each firm's "type," or for any given firm to know other firms' types. In addition, to the extent that farmers do not fully know the relationship between their input choices and ambient pollution, firms may not even know their own types, at least not regarding all components of the relevant vector. That is, while firms may have some knowledge of the cost of specific abatement activites, such as changes in tillage practices, they may have imperfect knowledge of exactly how much abatement takes place, and even less knowledge of the effect of these activites on environmental quality. These distinctions are important in modeling the relative amounts of information that agents have and what is presumed by a given regulatory strategy.

2.1.4. Stochastic Influences

Most NPS pollution problems, including agricultural NPS pollution, involve random variation or stochastic influences in production. It is important to distinguish between stochastic influences that enter into the system and uncertainty about the forms of the relationships that appear in the system. The distinction arises from the ability to learn over time regarding the latter. In this section we will discuss only the first type of uncertainty. Imperfect information about the relevant relationships is discussed below. In addition, firms may face stochastic input and output prices. The implications of this type of uncertainty for production decisions have been studied in detail. Since it is not an essential feature of the NPS pollution setting, we do not discuss this type of randomness here. However, since alteration of risk markets will alter input choices and hence pollution, these potential influences should not be ignored.

Stochastic production arises because the production process itself, both for crops and for environmental outputs, involves random natural influences, such as weather. Clearly, the actual yields resulting from given input levels depend on the realized rainfall and temperatures, which are not known at the time that (some) input choices are made. Likewise, the fraction of an applied pesticide that ultimately reaches a nearby water body will depend upon the amount and intensity of rain and wind in the period following application.

It is reasonable to separate the randomness in production into a component regarding crop production and one regarding environmental effects. There are at least partial risk markets for crops (crop insurance) and crop/income diversification strategies that help farmers in the management of production-related risks. However, for risks related to environmental impacts, such markets are not available. Any state-contingency must be built into the pollution policies under consideration.

2.2. *Firms*

Objectives and Behavior

In general, we may assume that farmers make choices as if to maximize the expected utility of their own profits. It may also be reasonable to suppose that farmers value environmental quality. For example, if a farmer's own well is contaminated by his use of agricultural chemicals, then any change in contamination levels will directly affect the farmer himself. Alternatively, farmers might use local water bodies for recreation. This suggests that environmental quality should be included as an argument in the farmer's utility function. Including environmental quality in the utility function of firm managers alters to some extent the details of the environmental control problem. In particular, there will exist more scope for voluntary emissions control when farmers care about environmental quality than when they do not.

The possibility of cooperative behavior may also exist. In the literature on common property resources it has been stressed that a variety of institutions may arise that lead to the optimality or near optimality of decentralized behavior, even without sole-ownership property rights. For example, in addition to profits, "guilt" or "community spiritedness" may enter into utility functions. In this case, pollution control "enforcement" mechanisms may exist within a group (without the need for outside enforcement), especially when the number of actors is fairly small (so free-rider problems are reduced).

Much of the literature assumes risk neutrality. However, it is well recognized that farmers exhibit risk aversion in many of their choices. Since contingent markets appear to be incomplete in agriculture, the failure to incorporate risk aversion may lead to erroneous predictions about how farmers will respond to environmental policies. In addition, the ability of some incentive schemes to induce efficiency hinges on firms' risk preferences. Random fines of the type discussed by Xepapadeas (1991), for example, can yield a Nash equilibrium if firms are risk averse and not risk neutral (Herriges *et al.*, forthcoming).

Choices

Most of the literature on NPS control assumes that firm managers choose either a single input, or one productive input and one that constitutes emission abatement. Alternatively, the firm can be viewed as choosing the level of a potentially polluting input (e.g., pesticide or fertilizer use). In practice, input choices and the resulting environmental quality are determined by a complicated set of factors, only a subset of which are policies designed specifically to control NPS pollution. For example, input and output prices, site characteristics (such as soil type and climate), degree of risk aversion, and supply-oriented government programs can all alter input use. These issues are usually ignored in models of NPS pollution. The simplified version of input choice used in most NPS models can be justified as a "reduced form" representation that subsumes the effects of these other policies and decisions. In this case, NPS policy choices that take as given other relevant elements of the policy choice set are "second best" in nature.

Similar considerations hold regarding the firm's output and risk management decisions. Farmers typically can choose portfolios of crops rather than just one. The chosen crop mix has important implications for environmental quality since some crops tend to be more pollution-intensive than others. Since crop mix is affected by agricultural policies, there are strong interactions between environmental and agricultural policies to consider. In addition, as risk management strategies, farmers can choose to hedge on future markets or participate in crop insurance schemes. Again, these choices may have important implications for environmental quality. For example, farmers with no form of crop insurance are thought to overuse pesticides as a precaution against the possibility of a devastating infestation (Antle, 1988). As with input use, most models of NPS pollution abstract from these details on the grounds that they can be subsumed into the relevant functions. However, in some cases the availability of risk management mechanisms may be important in evaluating NPS control policies, since some policies (such as ambient taxes) impose additional risks on farmers while others (such as regulation of management practices) do not.

In addition to choosing input and output levels, firms also can choose whether or not to remain in operation. Many policies aimed at curbing agricultural pollution have been accused of threatening the viability of some farm operations. Some NPS models explicitly consider the long run entry/exit decisions of firms and evaluate NPS policies in light of both their short run (marginal) and long run (infra-marginal or entry/exit) effects (Segerson, 1988). Of course, conditions ensuring both short run and long run efficiency are more stringent than those that ensure just short run efficiency.

In a non-cooperative setting where firms have private (asymmetric) information, the final set of choices that may be made concern any messages the farmer may send to other players. If there is a possibility of cooperation among farmers, then they might also have to decide whether to enter into a binding agreement with others to act cooperatively. While a cooperative solution may be preferred by the group as a whole, any individual firm may not perceive a benefit from the cooperative outcome and thus will have no incentive to enter into the agreement. Even in a non-cooperative setting, firms may be required or choose to report (not necessarily accurately) some information about themselves (e.g., their type). In addition, there may exist some sort of pre-play communication between farmers. Hence, information may be provided to other firms (especially if the number of farmers in a watershed is relatively small) and/or to the regulator.

2.2.1. Constraints

In making the above choices, firms are subject to a number of constraints. The most obvious constraint is the firm's technology.[2] As mentioned above, this may be conditioned on the actions of other agents, as when there are production externalities, or the emissions of several firms combine to affect environmental change. In the case of a stochastic technology, the constraints include the probability distribution function over outcomes of given actions, both of the agent and of other agents.

In addition to technological constraints, firms are also constrained by the actions of others and any possible response in such actions to those of the firm. In most cases, firms are assumed to hold Nash conjectures, i.e., they expect that other firms will not respond to changes in their own choices. Alternatively, the firm may hold non-Nash conjectures about the response of other agents to his or her actions (see Xepapadeas, 1992). These other agents include both the regulator and other farmers. In cases of non-Nash conjectures, the specific form of the conjectural variation must be specified, and the issue of consistency of conjectures in equilibirum must be addressed.

2.2.2. *Information*

In the NPS setting, the firms' decisions clearly depend upon its knowledge or expectations regarding its own technology or type. Given our focus in this paper, we will concentrate on the types and actions of agents and the level of environmental quality. Naturally, other relevant aspects of the information set exist as well. For example, the relationship between quasi-fixed factors and ex-post variable factors is an important determinant of the effect of risk on input demands. Similarly, the receipt of information relevant to choice of irreversible investments is an important determinant of investment decisions. These results are relevant since input choices determine NPS emissions. However, since these problems are not unique to NPS concerns, we will not address them further here.

As noted above, two aspects of type are relevant here. The first is the firm's utility or profit or cost function reflecting the costs of production and abatement. The second is the effect of the firm's input levels on environmental quality outside of the firm. Let the first type be given by θ^{π} and the second by θ^{ϵ}.

In most cases, the firm will know its own profit type, θ^{π}, which includes all information relevant to production and risk preferences. It may not, however, know with certainty the impact that its decisions will have on environmental quality. For example, it may not know the relevant fate and transport coefficients. Likewise, if there are interdependencies in either the production of outputs or the production of pollution, a firm's choices may also depend on the technologies and/or actions of other firms, which it may not know.

The importance of this lack of knowledge for the firms' decisions depends upon the type of policy used to control NSP pollution. For example, with policies that hinge only on firm-specific decisions (such as regulation of input use or management practices), once the policy has been set, each firm does not have to consider its own pollution type or the types/actions of other firms since its own profits are independent of those types/actions. However, with policies based on ambient pollution levels (e.g., ambient taxes), each firm's profits will depend on ambient pollution, which is in turn a function not only of its own type/actions but also of the types/actions of other firms. If true types are not known, it may be possible for an announcement to be made by the regulator regarding some aspects of types. For example, given an action by each farmer and realization of random natural events, it may not be possible to forecast exactly the resulting level of environmental quality. Thus, true types (production functions) are not known. However, the regulator may have on hand a model, albeit imperfect, for making such predictions. The regulator may announce that this model will be used to set policy parameters (e.g., tax rates or restrictions on input use), and the farmer may know this at the time choices are made. An analysis based on the use of a simulation model is described in Dosi and Moretto (1992). Thus, with ambient-based policies, each firm must consider other firms in making its own decisions. The Nash equilibirum notion is based on the ability of each firm to compute equilibria, and hence to base decisions on knowledge of the payoffs of all the firms.

Finally, at the time that production decisions are made, the firm does not know the values of the random variables that will determine output (including pollution). As with information about the pollution-generation function, the specific pollution control policy used will determine whether information about the random variables affecting pollution is relevant for the firm's decisions. In most cases, all firms are assumed to have the same beliefs regarding the probability functions and those beliefs are assumed to be correct. Hence, there is no problem of imposing consistency conditions regarding beliefs across agents. However, there is no particularly good reason for homogeneous beliefs to arise in this setting.

2.3. *Regulators*

2.3.1. *Objectives and Behavior*

In much of the regulation literature it is supposed that the regulator wishes to achieve efficiency in the economy by maximizing some measure of surplus of all agents in the economy (including both "firms" and "victims" of pollution). It is most often supposed that regulators seek to maximize the expected value of social surplus. However, it has been widely noted in the benefit-cost literature that this may not be a good welfare measure under uncertainty (e.g., Graham 1981). This issue has not been addressed in the NPS literature. Alternatively, it can be supposed that the regulator wishes to maximize his or her own utility function, which may or may not be directly and solely dependent on the wellbeing of agents. In this latter circumstance, models of bureaucratic behavior and rent-seeking activities may become relevant. There has been relatively little formal analysis of political concerns in the environmental economics literature. Magat *et al.* (1986) apply Peltzman's model of regulation to environmental policy. In their model, the pollution control agency sets regulations to maximize political support for the regulation. In this approach, the support of an individual or group either exists or it does not. Chen *et al.* (1993) allow the degree of regulation to be affected by "lobbying" contributions of groups in the economy, where lobbying effort is determined as the outcome of a non-cooperative game between these groups. They show that there is a difference between price and quantity instruments in the extent to which they give incentives to lobby. A recent paper by Laffont and Tirole (1991) analyzes the regulatory implications of possible "capture" between firms and the regulatory agency for the design of regulatory processes by legislation.

Given that the regulator's objectives may reflect broad social concerns, it may be that he or she includes in the objective function some sort of distributional weight between affected parties, or the social benefits of tax collection from pollution taxes (Dosi and Moretto, 1990). Presumably, the latter arise because of second-best considerations: the available commodity and income taxes are distorting, and replacement of these with pollution taxes carries some benefit from reducing deadweight losses. As is well-known in the general literature on incentive mechanisms, the incorporation of a social cost of raising funds can alter substantially the conclusions reached (Caillaud *et al.*, 1988).

Finally, it is recognized that different policies to control NPS pollution involve different transaction costs associated with monitoring and enforcement. In most studies of NPS pollution, these costs have not been explicitly incorporated into the choice among policies except to the extent that they change the social cost of raising funds, as just discussed. However, if there are substantial differences in transactions costs across policies, then measures of surplus should reflect these. One context in which transaction costs have been explicitly modeled is the choice between alternative rules for assigning legal liability for damages. For example, it is recognized there that different rules imply different incentives to sue and thus different total litigation costs. These costs have been explicitly considered (among other things) in the choice among alternative rules; see Shavell (1982).

2.3.2. *Choices*

Regulators can choose from among a number of alternative approaches to controlling pollution. The applicability of these alternative approaches depends on the particular NPS problem of interest. These include (1) policies directed to inputs, such as taxes on fertilizer or pesticides, or management practice requirements and (2) policies directed to outputs, such as output taxes or supply restrictions. For example, the Conservation Reserve Program used in the United States took certain land out of production partly to enhance water quality. Included here would be taxes on the activity in general, or zoning approaches. Additional possible policies are (3) taxes based on variations in ambient pollution levels, and (4) legal liability approaches. For example, legal liability might be appropriate in some cases of groundwater contamination but is not likely to be practical in the context of surface water. Nonetheless, much can be learned about surface water problems from the literature on liability. For example, the NPS setting described above is analytically similar to the case of "joint torts", which has been analyzed in detail in the law and economics literature (e.g., Miceli and Segerson, 1991). In fact, ambient taxes are analogous to joint liability for contamination.

In addition, while by definition NPS pollution generally rules out use of emissions-based policies, it may be possible to base policies on estimated emissions, as determined by a biophysical model of fate and transport relating inputs to emissions and ambient concentrations. This approach requires that the principal know or can find out firm types. This "indirect method" of controlling NPS pollution was originally suggested by Griffin and Bromley (1982). If emissions are perfectly correlated with an observable variable, use of the indirect method allows the design of first best policies (Griffin and Bromley, 1982). Even if the correlation between emissions and inputs is not perfect, if the model between the two is granted some sort of "policy legitimacy" and use of expected emissions based on the model is able to withstand legal scrutiny, then the indirect approach can be used. The implications of granting legitimacy to such models in NPS control policy has been studied by Shortle and Dunn (1986) and by Dosi and Moretto (1991).

In analyzing the NPS control problem, some studies have focused on a particular type of policy (e.g., ambient taxes), with the goal of determining whether the policy approach can be designed to ensure efficient incentives (e.g., Segerson, 1988; Xepapadeas, 1991, 1992). Other studies have tried to compare alternative policies to determine the conditions under which one approach is preferred to another (e.g., Shortle, 1984).

In addition to choosing a policy approach and specific policy parameters, in the presence of uncertainty, the regulator may also be able to make decisions regarding the generation or gathering of information about relevant parameters or relationships. In the complicated setting that we envision for the NPS pollution control problem, there are multiple aspects of the problem about which the regulator may learn. The possibility of learning over time is discussed in more detail below.

Finally, related to information generation is the regulator's decisions regarding monitoring. In general, the regulator can decide which variables to monitor (for example, firm management practices vs. ambient pollution levels) and the density of the monitoring (for example, how often and which locations). Most existing NPS pollution studies treat monitoring as exogenous. In practice, however, it will be endogenous. Since monitoring is an important determinant of the effectiveness of any policy and the associated enforcement costs, explicit monitoring needs/decisions should be recognized. There is a growing literature on the economics of monitoring and enforcing environmental regulations. The context, however, is point source pollution, where a regulator randomly audits or monitors and then imposes a penalty for non-compliance. Recently, Ellis (1992a, 1992b) has explored the interactions between monitoring decisions and the design of incentive mechanisms. The consideration of costly information and monitoring provides a bridge between the point and nonpoint literatures.

2.3.3. Constraints

The choice set of the regulator is constrained by prevailing institutions, especially legal ones. An obvious constraint is the inability to impose certain types of policies that distinguish among heterogeneous agents. With heterogeneous firms, efficient policy parameters (e.g., tax rates) would generally vary across firms. These may, however, be viewed as discriminatory, especially if the regulator is not able to prove with certainty the differential contributions of the firms. It may be viewed as arbitrary to base penalties on expectations regarding types, rather than actual types.

The existence of uncertainty may undermine the legal viability of other policies as well. For example, the ability to use liability approaches when there are multiple parties or a great deal of uncertainty may be constrained because of the difficulty of proving causation.

Finally, regarding incentive schemes, some studies impose a requirement that the regulator achieve a "balanced budget" from the scheme. While the justification for this is somewhat unclear,[3] this restriction has important implications for the feasibility of alternative schemes. In addition, it matters whether one imposes budget-balancing globally (for all outcomes) or only locally (at the efficient or equilibrium outcome). For example, Holmstrom (1982) has shown that under risk neutrality no globally budget-balancing mechanisms yields the efficient outcome as a Nash equilibrium even though if firms are sufficiently risk averse an efficient outcome is possible (Rasmusen, 1987). However, locally budget-balancing mechanisms can always be designed by combining full marginal payments with a set of lump sum rebates. In addition, in the absence of synergisms, such mechanisms can ensure efficient long run as well as short run decisions (Segerson, 1988; Miceli and Segerson, 1991).

2.3.4. Information

An essential feature governing the regulator's choice problem is the amount of information she has available about the types and actions (decisions) of the firms. In some of the NPS literature, actions are assumed to be known (observable) by the regulator but types are not (Dosi and Moretto, 1990; Shortle and Dunn, 1986). Hence, there is adverse selection but not moral hazard. In other studies, type is known but actions are not, implying moral hazard but not adverse selection (Segerson, 1988; Dosi and Moretto, 1990, 1991; Xepapadeas, 1991, 1992). With moral hazard, policies must be designed to induce efficient decisions by individual firms since these decisions are by assumption unobservable and thus cannot be directly controlled. Examples of such actions include the care taken by agents in the use of chemical inputs and the timing of fertilizer and pesticide applications.

With adverse selection, the regulator cannot predict a firm's behavior or responses to policies with accuracy. This assumes there is asymmetric information regarding production type. There are several aspects of environmental type which may involve private information for firms as well. For example, firms may be aware of small variations in type within a field (a slope change close to a stream for example) which the principal does not know. Other aspects of the environmental type may be either known or unknown to an equal degree by both the principal and the agent. In this case, there may be gains to having the principal conduct research and provide results to the agents. In a dynamic setting it may also be possible for the principal to infer private information (type) from a series of observations on input choices and public information regarding realized ambient pollution. For example, if the regulator imposes a tax or offers a subsidy, he or she does not know the magnitude of abatement response and hence the impact on environmental quality. One possibility in this circumstance is to have the firms send messages to the regulator regarding their type. It is possible to design incentive mechanisms under which it is in the firm's interests to report its true type to the regulator.

Alternatively, the regulator can simply set policy parameters on the basis of expectations regarding type, recognizing that the actual response will differ from the expected response. Following Weitzman (1974), it can be shown that in such cases different policy instruments have different implications for ex-post efficiency. With a flat supply curve and steep demand curve, a quantity instrument will minimize ex-post losses, while such losses will be minimized by a price instrument with a flat demand curve and steep supply curve. Empirical research on abatement costs for agricultural pollutants to surface water shows that the supply curve is flat and then becomes steeper with increasing abatement (Braden *et al.*, 1989). Thus, to achieve a low level of abatement (i.e. in areas with a low demand for water quality), a quantity instrument is preferred, while at high levels of abatement, a price instrument will be preferred, unless the demand curve is very steep. An application of Weitzman's analysis to the case of NPS is provided by Shortle (1984).

Even if the regulator knows firm types and actions, she may not be able to predict ambient pollution with certainty, due to both the influence of stochastic factors such as weather and imperfect information about fate and transport. In most studies, these two sources of uncertainty are combined while in others they are considered separately. The need to distinguish between them stems from possible differences in the information asymmetries and the ability to learn over time about fate and transport. Both the regulator and the firms are usually assumed to know the true probability distribution of the stochastic influences so there is no information asymmetry regarding these factors. However, the fate and transport coefficients for a particular type may be better known by the regulator than by the firm, if the regulator has greater access to scientific data or models.

2.4. *Dynamics*

2.4.1. *Stock Effects*

Decisions made by agents in the NPS setting have two types of effects on stock resources. One concerns elements of an agent's type; the other concerns environmental damages.

Many aspects of type depend on stock resources. For example, today's actions alter the rate of erosion, which alters future soil characteristics. This implies that agents' types will be changing over time at an endogenous rate (see Dosi and Moretto, 1991).

Moreover, some of the impacts of environmental quality also depend on intertemporal influences, such as the filling up of reservoirs with soil or the accumulation of nitrogen in water bodies. In this case, environmental harm may depend on both the flow rate of emissions as well as accumulated past emissions (Xepapadeas, 1991, 1992). This considerably complicates analysis of the pollution problem.

2.4.2. *Repeated Play*

When play is repeated and the regulation game becomes dynamic, a host of issues arise (see, e.g., Laffont and Tirole, 1988; Caillaud *et al.*, 1988). In such games, there are familiar problems of specifying the type of strategy used by the agents and the principal. For example, it is necessary to specify the regulator's ability to alter pollution policies once they are implemented, i.e., the use of an open loop versus closed loop strategy. "Once and for all" policies have very different properties than do policies that may be revised continually, especially if this revision potential (and the mechanism for revision) is public information. This concern is especially important when the revision depends in some manner on the actions of agents. While there has been considerable study of these issues in the general literature on regulation, and on application to the design of environmental policy (Malik, 1991), application to the NPS context is limited to Xepapadeas (1992), who compares open and closed loop strategies. Questions of the timing of the introduction of the regulation may also arise (Dosi and Moretto, 1991).

In situations of repeated play, there also may exist strategies that the farmers and/or regulator could use to achieve payoffs not obtainable in equilibirum in the static situation. However, to our knowledge, such issues have not yet been addressed in the context of NPS pollution.

2.4.3. *Learning*

When there is the ability to learn about other players' types, questions arise of how agents learn, and, if they are assumed to learn in a Bayesian manner, about the convergence of Bayesian equilibrium. It would also seem that in a repeated version of the moral hazard game with public information, the play between agents may be modified.

For these types of uncertainties, agents know that understanding of the relationships will evolve over time, possibly at an endogenous rate. This raises two questions: the ability to experiment, and the relationship between actions taken today and the arrival of information in the future. The literature on dynamic "option value" and the "irreversibility effect" (see, e.g. Hanemann, 1989) shows that the set of actions that are rational when learning is anticipated are different from those when it is not. The implications of this for repeated moral hazard and or adverse selection games have not been explored to our knowledge.

3. A FORMAL MODEL OF NPS POLLUTION REGULATION

In this section we formalize the discussion above in a model of the NPS pollution control problem. We state the model in continuous time because most of the existing dynamic research has been conducted in this framework. It is the simplest model that incorporates most aspects of the problem that have been studied in the literature. However, it is still enormously complex; we are not recommending actual investigation of this model. Hence, we will not address any of its features in detail nor place restrictions on it that may actually make its analysis tractable. This is the subject of the more specific investigations we will discuss in the next section of the paper.

3.1. *Technologies*

There are three firms in the model, indexed by $i \in I = \{1,2,3\}$. Firms one and two exist in the same "watershed." Firm 1 is "highest" in the watershed. Thus, firm 1's emissions affect firm 2's production function, but 2's emissions do not affect 1. Firms in $\{1,2\}$ and firm 3 are independent in this regard.

Each firm has a production technology giving output, $y^i(t)$ as a function of an input vector $x^i(t) \in \boldsymbol{R}^n$, a level of abatement effort, $a^i(t)$, production type,

$\theta^{\pi i}(t)$, and a random variable affecting crop production, $\mu^{\pi i}(t)$. If $i = 2$, then we also include the emissions for firm 1, given by $e^1(t)$. Thus, we have

$$y^i(t) = \begin{cases} f^i(x^i(t), a^i(t), \theta^{\pi i}(t), \mu^{\pi i}(t)), & i = 1,3 \\ f^i(x^i(t), a^i(t), \theta^{\pi i}(t), \mu^{\pi i}(t), e^1(t)), & i = 2 \end{cases} \tag{1}$$

It is assumed that each f^i has desirable properties, convenient to the specific model being studied.

Let p^i denote the random price of crops for firm i, let w^i be the vector of prices corresponding to x, and let $c^i(a)$ be the cost of abatement effort. These are distinguished by agent since firms may be spatially heterogeneous. Let $\tau^i(R(t))$ be the regulator's policy function (e.g., any taxes or subsidies relating to pollution). τ is a function of a vector R, which includes observed variables such as messages from firms, ambient pollution levels, or firm choices. While τ^i may be deterministic from the standpoint of the firm, it will generally be random. More will be said about $\tau(.)$ below. Since we wish to include messages from the firm, let $m^i(t)$ be the message sent at date t by firm i.

The firm's random profit at date t is given by

$$\pi^i(t) = p^i y^i(t) - w^i x^i(t) - c^i(a^i(t)) - \tau^i(R(t)). \tag{2}$$

Henceforth we will suppress the index for firm i when it will not cause confusion.

3.2. *Pollution*

Let firm emissions, $e^i(t)$, depend on the inputs it chooses, its abatement effort, its emissions type, denoted by θ^{ie}, and a random term μ^{ie}. Thus, we write

$$e^i(t) = \psi(x^i(t), a^i(t), \theta^{ie}, \mu^{ie}). \tag{3}$$

Note that the emission *type* of firm 2 may depend on the emissions of firm 1, (i.e. $\theta^{2e}(e^2)$) but we suppress this effect in the notation. This raises issues that have not been addressed in the NPS literature. One slight complication regards what happens when firms are not producing. If $x(t) = 0$, so the firm has exited the market, then emissions will be zero; hence, we have $e(0, ., ., .) = 0$ for all a and all realizations of the random variables. Any "background" level of emissions could be specified separately. Alternatively, one might wish to have emissions depend on how the farm was left at "abandonment," in which case emissions at $x = 0$ might not be zero for all types.

Pollution effects depend on the vector of emissions by all the firms. Let $e(t) = (e^1(t), ..., e^N(t))$, where $N(t)$ is the number of active firms at date t. The ambient pollution level at some receptor site, $P(t)$, is a state variable. If there are many receptors, then this must be an appropriately defined vector of states, and only the emissions and random terms that are relevant to that state need appear. The transition equation for $P(t)$ depends on emissions of all firms and stochastic natural events. Hence, we specify

$$dP(t) = \Phi(e(t), P(t))dt + \sigma^p P(t)dz^p(t), \tag{4}$$

where $\{z(t)\}$ is a Wiener process.

Note that this equation includes the vector $e(t)$, which generally will not be observed. In this case, the expected change in ambient pollution for a firm or the regulator, $E\{dP\}$, is not Φ itself, but the expected value of Φ, conditional on the firm's private information (if any). These expectations might arise from the use of some biophysical simulation model to translate management choices (x, a) into predicted emissions and pollution accumulation. Note too that we assume that ambient pollution is a known entity. If pollution is measured with some error process, a filtering problem arises; this raises well-known issues, which we do not address further.

Environmental damages, measured in dollars, are a random function of $P(t)$ and emissions. We denote these damages by $D(t) = \Gamma(P(t), e(t), \mu^D)$. The randomness could be associated with some random natural aspects of the relationship between emissions and environmental harm, or associated with the measurement of environmental damages in dollar terms, or both. The damage function specified here includes both the state variable P and the vector of the current flow of emissions, e. In this case, it may be that private decisions result in socially excessive abatement, rather than excessive pollution (Shortle and Miranowski, 1987). This is a result of both stocks and flows entering the objective function in a dynamic model. If the damage function is restricted to include only the pollution state variable, it easily can be shown for the Shortle and Miranowski model that private decisions lead to excessive emissions. The majority of the literature in which pollution accumulates places this restriction on the damage function.

3.3. *Type Dynamics*

There are two basic technological sources of dynamics in our model: the evolution of ambient pollution given in (3), and the evolution of the firm's type. The type transition equation is described by a stochastic differential equation of the form

$$d\theta^j(t) = \phi(x(t), a(t), \theta^j(t))dt + \sigma^j \theta^j(t) dz^j(t), \qquad j = \pi, e. \qquad (5)$$

The transition equation on type may define the evolution of something that is not observed. For example, any given firm may not observe a different firm's type, and the regulator may not observe the type of any firm. In these cases, the approach in (5) does not make sense. Rather, we must specify in some fashion the manner in which beliefs about type evolve, so that beliefs are a state variable of the system. One way to do this is to specify a model in which only the expectations of types are relevant, and to consider equation (5) as applying to the expectation about type that the firms or regulators may hold, conditional on any private information. Note that we would then have to introduce a different equation for each firm in the model, plus one for the regulator, for each firm's type.

3.4. *The Firm's Problem*

A time path of choices $\alpha^i = \{x^i(t), a^i(t), m^i(t)\}_t$ is called a plan. The payoffs to a firm depend on: i) the chosen plan, ii) the plans of other firms except i, α^{-i}, iii) the vector of types of the firms, (θ^i, θ^{-i}) (where θ^i is a vector of both the π and e-types), and iv) the time path of realizations of all the random elements of the system. Let Σ denote such a time path. Note that we do not include environmental quality here, although it could be an argument of the utility function, as discussed previously. The payoff from implementing the plan α^i when Σ occurs, types are θ, and the firm's opponents are playing α^{-i} is given by

$$B^i(\alpha^i, \alpha^{-i}, \theta^i, \theta^{-i}, \Sigma) = \int_t u^i(\pi^i(t))e^{-\rho t}dt, \qquad (6)$$

where u^i is the instantaneous utility function of firm i and ρ is the discount rate. Thus, the benefit of a plan is given by the present value of the utility of profits it generates under the realization Σ and the actions of others. We assume that each firm chooses its plan to maximize expected benefit, subject to constraints. Note that this specification allows firms to be risk averse.

The constraints faced by the firm include the policy choices of the regulator. For example, the regulator might require that some subset of choices for the firm lie in a set MP_i of input choices known as a "management practice." The practices are indexed by i since they will differ across firms, perhaps by the firm's type. Alternatively, the regulator might specify a set of emission standards (see further discussion below). Here, we treat management practices.

The firm's maximization problem is

$$P^i: \quad \max_{\alpha^i} E\ \{B^i(\alpha^i, \alpha^{-i}, \Sigma)|\Omega(0)\} \tag{7}$$

$$s.t.$$

$$dP(t) = \Phi(e(t), P(t))dt + \sigma^p P(t)dz^p(t), \tag{8}$$

$$d\theta^j(t) = \phi(x(t), a(t), \theta^j(t))dt + \sigma^j \theta^j(t)dz^j(t), \quad j = \pi, e \tag{9}$$

$$\tau = \tau(R(t)), \tag{10}$$

$$\alpha^{\theta i} \in MP^i. \tag{11}$$

(Note that one feasible action is to set $x = a = 0$ and go out of business.) The conditional expectation here is taken with respect to the joint distribution of all of the random terms defined so far, as well as the firm's beliefs regarding the actions and types of others, conditional on the firm's information set. The initial information set may include some elements of the firm's own type. Actual implementation of this model would require more detail about matters such as this joint distribution.

We have not employed any reaction function on the part of firms other than i when we specified firm i's problem. Hence, we are imposing Nash conjectures on the firm. We could alternatively include a conjectural variation term, but do not do so here, since most of the NPS papers to date have examined Nash equilibria.

One approach to solving the firm's problem is to suppose the use of feedback strategies and to employ dynamic programming. Alternatively, the maximum principle, generalized to incorporate uncertainty, may be employed. In the former circumstance, to derive Bellman's equation for this problem, let $S(t)$ be the state vector $(\theta^\pi(t), \theta^e(t), P(t))$ and $V(S(t))$ be the value function for P^i from date t onwards. Then we have

$$\rho V(S(t)) = \max_{\alpha i} E\{u(\pi^i) + \nabla V_S(.)\Upsilon + (1/2)[\sigma S]\nabla^2 V(.)[\sigma S]|\Omega(t)\}, \tag{12}$$

where Υ is a three-element vector function consisting of the transition equations in (8) and (9), σS is an obviously defined vector, and $\Omega(t)$ is the firm's information at t. Conformability of the matrices involved is assumed.

The information set Ω is the history up to t of all the variables assumed to be observable in the particular model being studied. Minimally, it includes the actions of the firm, the firm's private information about its type, ambient environmental quality, taken to be public information, and the time path of regulatory decisions. It may or may not include some of the random elements of interest; that is, some of the μ^j may be known when some of the x_i are chosen. In addition, some NPS models assume that the types or actions of other firms enter the information set.

One issue that we will not incorporate is the possibility of "experimentation" by the firm, in which the ability to alter the history, and hence future information, is recognized and acted on by the firm. If any learning is going on in the model we have specified, it is of a passive type, where history conditions the expectations, but information is not actively generated. The application of learning to dynamic NPS models has not been undertaken.

3.5. The Regulator's Problem

The regulator must choose a time path of pollution control policy functions, $\{\tau, MP\}_t$ based on some observable entities. We suppose that the regulator wishes to maximize the present value of some function of the economic surplus of consumers, profits of firms, environmental damages, the net tax revenues collected (if any) and transactions costs. A pollution policy will be called optimal if it maximizes the regulator's objective function, whether or not this corresponds to some accepted definition of optimality for society as a whole, such as Pareto efficiency.

The instruments that have been assessed in the literature include specification of input taxes, management practices, policies based on ambient environmental quality, policies based on emissions computed from observable variables and some biophysical model, joint liability rules, or some combination of these instruments. The policies may be deterministic or stochastic from the point of view of the firm.

Regarding input taxes, let the observable subset of inputs be given by $x^o \in \boldsymbol{R}^s$ where $s \leq n$. An input tax vector τ_x , differentiated by firm, can be applied to these inputs.

*MP*s are represented by constraints on the input combinations that a farmer can use (see (11) above). These input-based policies might be differentiated by some aspects of firm type, but it typically is the case that *MP*s are not fully heterogeneous. The elements of type that are used to specify *MP*s could be all of those known to the regulator or a subset thereof. If an emission standard is used instead of a *MP* requirement, the standard could be made equivalent to constraint (11) in the firm's problem by setting *MP* equal to the set of inputs such that emissions do not violate the emission standard.

Ambient pollution taxes are very similar to strict liability rules based on notions of joint liability, especially when the probabilities of winning or losing litigation are ignored. Therefore, we only treat the pollution tax case here (see Miceli and Segerson, 1991, for the joint liability case). Let the vector τ^p (again, one element for each firm) denote this ambient tax scheme. If budget-balancing is required and firms are risk averse, the same outcome can be achieved by a system of random fines, in the spirit of Xepapadeas (1991), as will be discussed in more detail below.

The information available to the regulator at date t, denoted by $\Omega^R(t)$, consists of, at a minimum, the history up until t of observed inputs used on the farm, messages received from them, and ambient pollution levels:

$$\{x^o(s), m(s), P(s)\}_{s\leq t} \subseteq \Omega^R(t). \tag{13}$$

Here, the observed inputs and messages represent vectors with an element for each firm. Variations in the information set include adding firms' types, or assuming that all inputs are observable.

If type is known, either because the regulator observes it or an incentive mechanism has been implemented under which reporting true type is in the interests of the firm, then the regulator might predict emission and/or pollution levels based on a mathematical model, and then apply an incentive scheme to inputs at the farm. This is called the indirect approach in the NPS literature. Alternatively, it might observe ambient pollution levels, and base its interventions on this information. This is known as the direct approach.

Given an information set at t, the regulator chooses its incentives. Hence, we may write $R = R(\Omega^R(t))$. As mentioned, we assume that these are chosen to maximize the expected present value of the regulator's utility. Let $U^R(C, \Gamma, \pi, \tau)$ denote this utility. It depends on consumer surplus for crops (C), the level of environmental damages, the profits of firms, and the vector of taxes or subsidies paid, including any transactions costs. Note that the regulator's problems may explicitly incorporate price effects in output and/or input markets, while the firms' problems do not under an assumption of competition.

Naturally, the realized utility from the time path of the pollution control policies, $\{\tau(.), MP(.)\}_t$, depends on the realization of random variables affecting pollution and profits. It also depends on the choices made by the firms when the mechanism is $\{\tau(.), MP\}_t$. These choices depend in turn on the kind of equilibrium one is supposed to have when choices are made by firms. Let $\{\alpha(\{\tau, MP\}, E)\}$ denote the time path of choices by firms when the regulator uses the time path of policies $\{\tau, MP\}$ and choices arise from an E-equilibrium. Then U^R can be evaluated at each date. Let $B^R(\alpha, E, \Sigma)$ be the integral of U^R under a realization of stochastic terms Σ. Then the regulator solves

$$
\begin{aligned}
P^R: \qquad & \max \; E\ \{B^R(\alpha, E, \Sigma)|\Omega^R(0)\} \\
& s.t. \\
dP(t) &= \Phi(e(t), P(t))dt + \sigma^p P(t)dz^p(t), & (14)\\
d\theta^j(t) &= \phi(x(t), a(t), \theta^j(t))dt + \sigma^j\theta^j(t)dz^j(t), j = \pi, e, & (15)\\
(\tau, MP) &\in L, & (16)
\end{aligned}
$$

where L is the domain of policy choice. It reflects any legal or institutional restrictions that might impinge on the choice of policy instruments.

The same caveats about the transition equations on pollution and types regarding observability arises here as it did for firms. Expectations in P^R are taken with respect to the regulator's information. Unless some assumptions are placed on the relationship between the regulator's information and the firms' information, conceptual difficulties of the infinite regress type arise. Thus, the regulator's problem requires a guess about what firms' beliefs are to compute firms' choices, but the latter depend on what the firm believes the regulator believes, etc.. In the NPS work we will review, such assumptions as homogeneous beliefs are typically imposed. Note that these difficulties may not arise if we are dealing with a simple form of Nash equilibrium, which only requires "consistency" of beliefs; however, they may arise when various equilibrium refinements are considered.

In addition to the controls identified in P^R, the regulator might also invest in information to reduce uncertainty. There are two basic objectives that such efforts might be directed to: increased monitoring so that unobserved entities are observed, and scientific research so that unknown relationships are better specified. We leave the formal specification of such activities out of our "simple" model.

4. THE NPS LITERATURE

In this section we provide a brief overview of existing work on the use of incentive mechanisms to control NPS pollution. There are obvious problems in delineating the research that comprises the NPS literture. There is a large body of literature on the economics of regulation in general that is of relevance (see Laffont and Tirole, 1993 for a thorough review). However, we limit our discussion to those efforts that explicitly address NPS problems, and, within the NPS literature, to those papers that address the implications of asymmetric information. These papers fall into two groups, those that deal with the adverse selection problem and those that instead address moral hazard.

4.1. Adverse Selection

Adverse selection arises when the principal cannot observe firm types. One goal of the policy intervention is to specify a contract between principal and agent such that agents provide truthful messages about their type. According to the revelation principle, messages can be restricted to "direct" ones, where type itself is reported.

Spulber (1989) analyzes such a contract, as do Dosi and Moretto (1990). In Spulber (really a point source paper), the firm has private information about the abatement cost function. There are many firms, and it is common knowledge that all participants hold beliefs that types are drawn from a single distribution function. Firms are risk neutral. The firm's aggregate output is large relative to the market. There is a conflict between achieving emission reduction and increasing prices. The regulator wants to maximize consumer surplus net of environmental damages and production costs, plus lump sum rebates of tax receipts. Since consumers are rebated all taxes, budget balance holds. Attention is given only to fully separating equilibria.

Spulber shows that the direct revelation game has truth telling as a Bayesian Nash equilibrium. The full information allocation can be attained if and only if the net benefits to consumers exceed the information rents that must be paid to firms in the incentive scheme. He goes on to show that both individual and aggregate output and emissions are lower under asymmetric information than under full information.

Dosi and Moretto (1990) specify a static model in which the relationship between inputs and emissions is conditioned on a single firm's type. This is θ^e

as specified above. The relationship between emission and pollution damages is deterministic. Type takes on one of two values, and the principal's belief regarding the probabilities are common knowledge. Both principal and agent are risk neutral. The principal's objective function takes the form of firm profit, less environmental damages, plus a positive parameter times tax revenue. The incentive mechanism takes the form of a tax and a management practice requirement after the message is received. The authors apply standard approaches to such problems to derive a mechanism under which truthful reporting of type occurs.

Shortle and Dunn (1986) suppose that the principal does not know the firm's profit type. The model is static. The emission function is uncertain both because of random rainfall and the fact that the principal does not fully know the form of the function. In a static model this distinction is not of interest, however. The principal can observe input choices. The regulator chooses incentive schemes to maximize expected profits less environmental damages, where the latter is either a known or random function of emissions. *MP* standards and incentives and estimated runoff standards and incentives are considered.

Shortle and Dunn show that under risk neutrality, when there is a single agent, an *MP* incentive scheme can be used to bring about *ex ante* efficient practices. The *ex ante* optimum cannot be achieved with *MP* standards. Regarding estimated emission policies, these can be used to achieve efficiency, but only if the damage function is non-linear.

In the case of multiple agents, Shortle and Dunn (1986) acknowledge the theoretical applicability of revelation schemes, but dismiss them as impractical. The authors note that when risk aversion is present, the same conclusions hold as under risk neutrality, as long as the principal knows the agent's utility function. This would have to be elicited via a revelation scheme.

4.2. Moral Hazard

Several authors have addressed the moral hazard issue, in which input choices are not observed. The first was Segerson (1988). In the Segerson paper there are many heterogeneous firms. Emissions are a deterministic function of inputs, but environmental damages are a stochastic function of all firms' emissions. Firms are risk neutral, and employ Nash conjectures. The profit functions, functions relating inputs and realizations of the random term to damages, and the objective distribution of the random term are common knowledge.

The regulator has all of the above information, except that she cannot monitor input usage. The regulator seeks to maximize expected profits less expected damages. There is no requirement of budget balancing. The principal observes ambient environmental quality, and bases an incentive scheme on monitored pollution levels. The model is static, but considers both short run and long run equilibria.

Segerson derives an incentive scheme similar to that proposed by Holstrom (1982) for problems of moral hazard in teams. Correct short run marginal incentives can be achieved when each firm that is active in equilibrium (i.e. uses the polluting input) pays the full marginal damages, rather than just that firm's share of the damages. This scheme avoids free-riding, since each firm is faced with all of the impact of its decisions. If the damage function is linear, then the appropriate ambient quality tax is marginal damages. Since the full marginal damages are paid by each firm, the total tax collected on the marginal unit of pollution exceeds the damages from that unit. Hence, the scheme does not achieve budget balance at the margin.

To ensure correct long run incentives, each active party should receive a reduction in tax payment equal to the expected damages that would have arisen had the firm been inactive (either not farming or not using the polluting input). Given this tax credit, the total payments of each firm reflect its incremental contribution to expected pollutant loadings. The incremental loadings are those with the firm operating at an optimal short run configuration, less loadings without the firm.

The total payments summed over firms may be greater than, less than, or equal to total damages, depending on the shape of the damage function. Budget balance will hold in equilibrium if polluters are independent in causing damages. In our example, budget balance holds if

$$\begin{aligned} E\{\Gamma(P,e^1,e^2,e^3,\mu^D)\} &= E\{\Gamma(P,e^1,0,0,\mu^D)\} \\ &+ E\{\Gamma(P,0,e^2,0,\mu^D)\} \\ &+ E\{\Gamma(P,0,0,e^3,\mu^D)\}. \end{aligned} \tag{17}$$

However, due to the presence of synergism in pollution effects, (17) is unlikely to occur in practice. In this case, while it will still be possible to budget balance through a set of lump sum subsidies, such an approach will not yield efficient long run incentives.

Moral hazard also has been studied in two papers by Dosi and Moretto (1990,

1992), one static and one dynamic, and in two dynamic papers by Xepapadeas (1991, 1992), one deterministic and one stochastic. We will discuss these in turn.

Dosi and Moretto (1990) study a static moral hazard problem with a single risk-averse agent, who does not have private information regarding type. The principal conditions an incentive scheme on ambient pollution levels. Pollution is a function of firm choices, the known type, and a random term. The principal seeks to maximize the expected value of the firm's net profits, minus environmental damages. There are no price effects in the model. It differs from those considered earlier by the inclusion of a benefit to the principal from collecting taxes. They show that the moral hazard problem leads, as expected, to an inefficient outcome relative to the full information situation. The tax with asymmetric information is greater (less) than the full-information tax if ambient pollution levels are such that ambient pollution divided by the likelihood ratio is non-negative (positive). The tax schedule is increasing in the ambient pollution level under the monotone likelihood ratio property. They also show that under risk neutrality of both principal and agent, a first-best allocation can be achieved. If the social value of tax receipts is zero, then the model is equivalent to a single-firm, short-run version of the paper by Segerson (1988).

In Dosi and Moretto (1991), the problem is somewhat different. There is a single, risk neutral firm. Although the firm's type is known, it evolves according to a stochastic differential equation similar to equation (5) above. There are two kinds of inputs: those related to production and "maintenance" inputs which alter the expected rate of change of type. These are separable in the profit function. The principal observes the former inputs but not the maintenance inputs. The law governing type evolution is common knowledge. Pollution and damages are a deterministic function of inputs and type. The paper therefore represents an application of the "indirect" approach discussed above. The main issue addressed in this paper is the time profile of environmental fees. In particular, attention is given to a possible lag before fees are introduced.

Xepapadeas (1991) presents a dynamic moral hazard model. The model is deterministic. Pollution accumulates according to a non-stochastic differential equation of the form in equation (4), but with $\sigma^p = 0$. There are several firms, each of which produces output using capital and labor. Capital and labor can also be allocated to abatement effort. Emissions are a function of output and abatement. The function is strongly separable between these two, so that abatement does not directly affect output. There is a representative consumer who derives utility from consumption of the single produced goods and environmental quality.

Xepapadeas first solves for the optimal time path of abatement and pollution by solving a social planner's problem whose objective is to maximize the present value of a utility integral over an infinite time horizon. The co-state variable on the pollution transition equation is the social shadow cost of pollution.

It then is supposed that the social planner cannot observe abatement, but can observe firm type (there are many identical firms) and ambient pollution levels. It is shown that under an efficient contract at each date the marginal value product of abatement (the co-state variable times the derivative of pollution concentration with respect to firm i's abatement holding other firms' abatement constant at their optimal abatement) will equal the marginal abatement cost.

Xepapadeas considers the efficiency of several contracts. The first includes a stochastic system of fines, based on those proposed by Rasmusen (1987) for moral hazard problems in teams. Under this scheme, if observed pollution equals the socially efficient level, then each firm receives a share of the net social benefit from optimal abatement. If pollution exceeds the optimal level, one firm, say firm i, is randomly chosen to pay a fine, while all others (not chosen) receive a subsidy. The scheme provides budget balancing. While Xepapadeas considered risk neutral firms, Herriges *et al.* (forthcoming) have shown that a sufficient degree of risk aversion is needed to ensure that a Nash equilibrium under this scheme will be efficient. Xepapadeas also sets forth incentive schemes closely related to those proposed by Segerson (1988). It is important to note that only short run incentives are considered, and the number of active firms in equilibrium does not arise.

While the model just described is a dynamic one, the issues that arise are not inherently dynamic. This is not the case in Xepapadeas' 1992 paper. There, a stochastic dynamic model is studied. There is only moral hazard, and no adverse selection. There are many firms, each with a known and constant type. Pollutants evolve according to a stochastic differential equation as in equation (4) above. Pollution damages are a deterministic function of pollution levels. Firms maximize the expected present value of profits by choosing a time path of emission levels. The social planner's problem is to maximize the expected present value of profits minus environmental damages. The planner cannot observe emissions, but can observe pollution concentrations.

Xepapadeas begins by establishing the relationship between several long run variables of interest. Let $P^{\pi}(ss)$ be the steady state pollution concentration under profit maximization in a deterministic setting and $E\{P^{\pi}(ss)\}$ be the expected value of the limiting distribution for pollution levels under profit maximization when there is uncertainty. Let $P^{*}(ss)$ and $E\{P^{*}(ss)\}$ be the corresponding variables for the social planner's problem. Xepapadeas' first result is that

$$P^{\pi}(ss) = E\{P^{\pi}(ss)\} > P^{*}(ss) > E\{P^{*}(ss)\}.$$

When turning to an incentive scheme, it is supposed that the regulator's objective is to devise a scheme such that the mean value of the limiting private pollution distribution is driven to the mean of the limiting socially optimal distribution. Xepapadeas considers two solution concepts: open loop (commitment) Nash equilibrium and feedback Nash equilibrium, which is subgame perfect.

Xepapadeas studies both deterministic and stochastic pollution dynamics for both equilibrium notions. Beginning with the deterministic case, the incentive scheme takes the form of a tax in which all firms pay the same amount equal to the social valuation of deviations of observed from optimal pollution levels, times the sum of the discount rate, and a depreciation factor for pollution accumulation, b. Hence,

$$\tau^{ol} = -\lambda(P - P^{*})(\rho + b).$$

For the feedback equilibrium, conjectures about emissions of others depend linearly only on a firm-specific constant and a constant, a, times pollution levels. In this case we have

$$\tau^{fb} = -\lambda(P - P^{*})(\rho + b + (n-1)a).$$

For both schemes, Xepapadeas shows that the pollution level converges to the socially optimal steady state pollution level. Also, as long as $a < 0$, then $\tau^{ol} < \tau^{fb}$.

If the static incentive scheme derived by Segerson is imposed instead of the dynamic schemes, then under open loop equilibria, pollution converges to a lower level than the optimum. The feedback equilibrium levels converge to a larger pollution stock than under open loop equilibrium. For the feedback equilibrium, the relationship to the optimal steady state pollution level is indeterminant.

Similar results are derived for the case of uncertainty, but now taxes depend as well on the uncertainty parameter σ. The analysis is only undertaken for the feedback Nash equilibrium. The mean of the pollution process under the incentive scheme converges to the mean of the pollution process governed by the social planner.

5. DIRECTIONS FOR FURTHER RESEARCH

In this section of the paper we provide some suggestions for future research needs in the theory of NPS pollution control. Certainly these suggestions are not exhaustive; there is much to be done. But they are areas that we believe are deserving of attention.

(1) *Active learning/social experiments*: For some of the information problems inherent in NPS pollution control there are possibilities for improving information (learning) over time. Learning may be passive or active.

Passive learning has not been addressed in any substantive fashion in the NPS literature. However, there exist more general economic analyses upon which such analyses could be based. For example, suppose that $\Gamma(.)$ embodies threshold effects, as mentioned above, but that the exact locations of the thresholds is unkown. Then at some time, the ambient pollution process may pass a threshold, and information will jump. The work on dynamic resource problems with uncertain technology might usefully be employed here (e.g. Dasgupta and Stiglitz, 1981; Desmukh and Pliska, 1985).

Regarding active learning, research aimed at designing learning experiments would be useful. This could involve the set up of "social experiments" whose main purpose is to generate information. These would involve deliberately manipulating the system in what may appear to be a sub-optimal way in order to improve our understanding of the relevant relationships. While such experiments may be politically unpopular, they might improve efficiency in the long run. From a research perspective, the question is the following. If one were to incorporate the possibility of learning into an NPS model, how would the policy choices today differ from those in a world where learning is not possible? For example, would it be sensible to adjust policy parameters (to levels that are sub-optimal from a myopic perspective) to generate information that can then be used to improve the choice of policy parameters at a later date? What form of experiment would one want to undertake?

(2) *Expectations/Interactions among firms*: The NPS research to date is based primarily on Nash assumptions about firms, i.e., firms are assumed to believe that other firms will not respond to changes in their own input and output levels (the exception being Xepapadeas, 1992). Clearly, this is inappropriate in a setting where there are only a small number of firms and firms have at least some

information about other firms' types (as would be expected at the watershed level where firms know their neighbors). The interactions among firms is an important consideration in the formulation of NPS policies. As noted above, it greatly complicates the analysis as well as raising possible legal issues such as the ability to base one firm's tax payments on the actions of other firms. The importance of firm interactions suggests that a more thorough analysis is needed of how firms interact and how their expectations regarding the behavior of other firms affect model results. Research directed at understanding both the nature and the implications of complicated firm-level interactions is needed. In particular, the possibility and characteristics of non-Nash equilibria need to be investigated.

Interactions among firms may also include technological interactions among a small group of nearby farms, e.g. our firms 1 and 2. The implications for NPS policies of Coasian bargaining between them should be addressed.

(3) *Disequilibrium Behavior*: In general, microeconomic analysis is focused on equilibrium behavior. The literature on NPS pollution control is no exception. However, given the information problems inherent in NPS control and the potentially complicated interactions among firms, the possibility of disequilibrium seems particularly relevant in this context. In addition, politicians are likely to be very interested in what would happen initially when a particular NPS policy is implemented, before an equilibrium is reached. For example, given initial expectations about other firms, how are firms likely to respond to policies initially? (This could determine the political acceptability of the policies.) How would the system then move to an equilibrium? Is the equilibrium stable?

A related question regards the non-uniqueness of equilibria. Two issues arise here: first, the well-known one regarding Nash equilibrium and how firms come to find themselves in a particular equilibrium (e.g., via evolutionary processes) and second, the possibility that some equilibria may be more attractive than others in some sense, especially in a dynamic model. Regarding the latter, suppose that we are interested in a Bayesian equilibrium in an adverse selection game. Truth-telling may be a rational strategy for a firm, given that others report truthfully. But there may exist equilibria where all firms report untruthfully. Moreover, this may be "enforced" in repeated play to the advantage of the farmers collectively. The ability of the regulator to discover this may depend on the availability of certain types of public information and the legality of certain types of regulations.

(4) *Multiple Instruments*: The information problems inherent in NPS pollution control suggest that no single instrument is likely to achieve a first-best outcome. It may be preferable to use a combination of instruments. In practice, this appears to be the approach that is being taken. Yet little theoretical research on the role of multiple instruments has been done. An exception is Braden and Segerson (1991). To date, most NPS research has analyzed single-instrument policies, without consideration of how alternative instruments might interact or complement each other. Thus, research that explicitly analyzes instrument packages seems to be needed. The relevant research question is how to design a package of instruments that is more effective than any single instrument alone in controlling NPS pollution.

(5) *Cooperative Behavior*: The NPS research to date assumes that firms' behavior is non-cooperative, regardless of the policy instrument chosen. In many settings, however, the number of firms (e.g., within a watershed) may be sufficiently small to allow cooperative behavior (e.g., water quality coops). Thus, research exploring the implications of possible cooperative behavior seems to be needed. This research should not simply explore the outcomes in a world where firms happen to want to behave cooperatively. Rather it should consider the design of mechanisms to foster cooperative behavior. For example, are there some policy instruments that would cause firms to want to behave cooperatively, i.e., are the incentives for cooperation different under different policy instruments? If so, a comparison of instruments needs to recognize not only their differential impacts on input/output choices but also on the choice regarding cooperative vs. non-cooperative behavior.

(6) *Enforcement Costs*: There has been increasing interest in and analysis of enforcement and monitoring issues in the context of point-source pollution (e.g., Russell *et al.*, 1986; Malik, 1991, etc.). However, the NPS literature to date largely ignores endogenous decisions regarding enforcement and monitoring. It simply assumes that some variables can be observed/monitored (at zero cost) and others cannot (or can only be observed at a prohibitive cost). In practice, the extent and nature of monitoring is endogenous. The regulator can decide how often to monitor and which sites to monitor. Monitoring determines both the extent and the quality of information available, which in turn determine the effectiveness of alternative policy instruments in controlling NPS pollution.

These considerations need to be introduced explicitly into NPS models and evaluation of policies should reflect differences in the associated enforcement /monitoring costs.

(7) *Dynamic Games*: The dynamic games studied to date take very simple forms. The more general game theory literature has studied regulation issues in a dynamic context and identified the potential importance of reputation effects. These issues should be explored in the NPS context.

6. CONCLUDING REMARKS

The problem of the control of NPS pollution is a challenging one, not just in applied policy analysis, but in theory as well. We have attempted to provide an overview of the difficulties that arise in the NPS setting, with emphasis on the theoretical modeling issues that must be faced. We also have attempted to provide a brief guide to the literature and to suggest some areas for further research.

It is our opinion that the NPS problem is a worthy one for research. First, the pollution concerns are of great importance. Especially in the developed countries, but also in the developing world, the existing environmental pollution problems are heavily tied to nonpoint sources, and these sources present the most vexing policy dilemmas. Second, the issues raised for the design of regulatory policy lie at the forefront of research in economics. The research directions listed above in the application of dynamic game theory to moral hazard and adverse selection problems are challenging ones in pure theory. The tie to real problems provided by applications to NPS may offer some structure for obtaining results in such investigations. Third, the policy considerations arising from complex interactions among firms and regulators and nature arise as well in other problems in public economics where externalities exist.

NOTES

1. Implicitly, "multiple" means more than one but not so many as to make treating the polluters as a team untenable. For example, cases involving millions of automobile drivers whose emissions combine to determine ambient air quality are not considered in the standard NPS literature.

2. While the most general results can be obtained by considering general forms of the relevant functions, in some cases, specific functional forms (e.g., quadratic) are chosen to simplify the analysis without loss of insight (e.g., Shortle, 1984).
3. The original literature on moral hazard in teams (Holmstrom, 1982; Rasmusen, 1987) was in the context of the organization of firms, where budget balancing was simply a (logical) constraint that all of the output be distributed among workers. It is not clear, however, that the logic carries over to the context of pollution control. One possible justification might be based on legal principles. For example, collection of total tax payments that exceed the value of pollution damages may be viewed as "double recovery". In the context of legal liability, even under joint and several liability (where multiple firms are potentially liable), plaintiffs suffering pollution damages are barred from recovering from the responsible parties a total amount in excess of the actual damages. However, punitive damages allow for payments that exceed damages. Likewise, a justification might be based on long run entry/exit incentives (see Segerson, 1988). Yet, the papers that require budget-balancing (e.g., Xepapadeas, 1991; Herriges *et al.* forthcoming) do not consider long run incentives. Given the pivotal role of the budget-balancing constraint, further justification seems warranted.

REFERENCES

Antle, J.M. 1988. *Pesticide Policy, Production Risk and Applied Welfare Analysis.* Washington, D.C.: Resources for the Future.

Braden, J., G. Johnson, A. Bouzaher, and D. Miltz. 1989. "Optimal Spatial Management of Agricultural Pollution." *American Journal of Agricultural Economics,* 89, pp. 404–413.

Braden, J. and K. Segerson. 1993. "Information Problems in the Design of Nonpoint-Source Pollution Policy." In C. Russell and J. Shogren, Eds. *Theory, Modeling, and Experience in The Management of Nonpoint-Source Pollution.* Kluwer Academic Publishers, Boston, pp. 1–35.

Caillaud, B., R. Guesnerie, P. Rey, and J. Tirole. 1988. "Government Intervention in Production and Incentives Theory: A Review of Recent Contributions." *Rand Journal of Economics* 19(1), pp. 1–26.

Chen, K., T. Tomasi, and T. Roe. 1993. "Political Economy and Pollution Regulation: Instrument Choice in a Lobbying Economy." Staff Paper, pp. 93–44. Michigan State University, Dept. of Agricultural Economics.

Dasgupta, P. and J. Stiglitz. 1981. "Resource Depletion under Technological Uncertainty." *Econometrica,* 49, pp. 85–104.

Desmukh, S. and S. Pliska. 1985. "A Martingale Characterization of the Price of a Non-renewable Resource with Decisions Involving Uncertainty." *Journal of Economic Theory,* 35, pp. 322–42.

Dosi, C. and M. Moretto. 1990. "Incentives for Nonpoint Source Water Pollution Control

Under Asymmetric Information," Mimeo, Dipartimento di Scienze Economiche, Universita di Padova, Italy.

Dosi, C. and M. Moretto. 1992. "Non Point Source Pollution, Maintenance Policy and the Choice of Time Profile for Environmental Fees." NOTA DI LAVORO 2.92. Fondazione ENI Enrico Mattei, Milan, Italy.

Ellis, G. M. 1992a. "Optimal Environmental Regulations with Asymmetric Information but Without Commitment to Monitoring." Unpublished manuscript. University of Washington.

Ellis, G. M. 1992b. "Enforcing Incentive Compatible Environmental Regulations." Unpublished Manuscript. University of Washington.

Fudenberg, D. and J. Tirole. 1991. *Game Theory*, Cambridge, MIT Press.

Graham, D. A. 1981. "Cost-Benefit Analysis Under Uncertainty," *American Economic Review*, 71(4), pp. 715–725.

Griffin, R. and D. Bromley. 1982. "Agricultural Runoff as a Nonpoint Externality." *American Journal of Agricultural Economics*, 64, pp. 547–52.

Hanemann, W.M. 1989. "Information and the Concept of Option Value." *Journal of Environmental Economics and Management*, 16, pp. 23–37.

Herriges, J., R. Govindasamy, and J. Shogren. Forthcoming. "Budget-balancing Incentive Mechanisms,"*Journal of Environmental Economics and Managment*.

Holstrom, B. 1982. "Moral Hazard in Teams." *Bell Journal of Economics*, 13, pp. 324–40.

Kozloff, K., K.W. Easter, T. Tomasi, I. Moore, S. Taff, and Y. Wang. 1990. "An Evaluation of Options for Micro-Targeting Acquisition of Cropping Rights to Reduce Nonpoint Source Water Pollution." Final Report to U.S. Environmental Protection Agency. Department of Agricultural and Applied Economics, University of Minnesota.

Laffont, J.J. and Tirole, J. 1988. "The Dynamics of Incentive Contracts." *Econometrica* 56, pp. 1153–1175.

Laffont, J.J. and J. Tirole. 1991. "The Politics of Government Decision-Making: A Theory of Regulatory Capture," *Quarterly Journal of Economics*, 106, pp. 1089–1127.

Laffont, J.J. 1992. "Regulation of Pollution with Asymmetric Information," in C. Dosi and T. Tomasi, Eds. *Nonpoint Source Pollution Regulation: Issues and Analysis*. This volume.

Laffont, J.J. and J. Tirole. 1993. *A Theory of Incentives in Procurement and Regulation*. Cambridge, MA: MIT Press.

Magat, W. A., A. J. Krupnick and W. Harrington. 1986. *Rules in the Making: A Statistical Analysis of Regulatory Agency Behavior*, Washington, D.C., Resources for the Future.

Malik, A. 1991. "Permanent versus Interim Regulations: A Game-Theoretic Analysis." *Journal of Environmental Economics and Management*, 21, pp. 127–139.

Miceli, T.J. and K. Segerson. 1991. "Joint Liability in Torts: Marginal and Infra-marginal Efficiency," *International Review of Law and Economics*, 11, pp. 235–249.

Park, W. M. and L. A. Shabman. 1982. "Distributional Constraints on Acceptance of Nonpoint Pollution Controls," *American Journal of Agricultural Economics*, 64(3), pp. 455–62

Peltzman, S. 1976. "Toward a more general theory of regulation," *Journal of Law and Economics*, 19, pp. 211–240.

Rasmusen, E. 1987. "Moral Hazard in Risk-Averse Teams." *Rand Journal of Economics*, 18, pp. 428–435.

Russell, C. S, W. Harrington and W.J. Vaughan 1986. *Enforcing Pollution Control Laws.* Washington D.C.: Resources for the Future.

Segerson, K. 1988. "Uncertainty and Incentives for Nonpoint Pollution Control," *Journal of Environmental Economics and Management*, 15, pp. 87–98.

Shavell, Steven. 1982. "The Social versus the Private Incentive to Bring Suit in a Costly Legal System." *Journal of Legal Studies*, 11, pp. 333–340.

Shortle, J. 1984. "The Use of Estimated Pollution Flows in Agriculture Pollution Control Policy: Implications for Abatement and Policy Instruments," *Northeast Journal of Agricultural and Resource Economics* 13, pp. 277–285

Shortle, J. and J. Dunn. 1986. "The Relative Efficiency of Agricultural Source Water Pollution Control Policies." *American Journal of Agricultural Economics*, 68, pp. 668–77.

Shortle, J. and J. Miranowski. 1987. "Intertemporal Soil Resource Use: Is It Socially Excessive?" *Journal of Environmental Economics and Management*, 14, pp. 99–111.

Spulber, D.F. 1985. "Effluent Regulation and Long-Run Optimality." *Journal of Environmental Economics and Management*, 13, pp. 103–16.

Spulber, D.F. 1989. "Optimal Environmental Regulation Under Asymmetric Information," *Journal of Public Economics*, 35, pp. 163–181.

Stoley, N. 1989. "Reputation and Time Consistency," *American Economic Review*, 79, pp. 134–139.

Weitzman, M. 1974. "Prices vs. Quantities." *Review of Economic Studies*, 41, pp. 477–91.

Xepapadeas, A. 1991. "Environmental Policy under Imperfect-Information: Incentives and Moral Hazard," *Journal of Environmental Economics and Management*, 20, pp. 113–126.

Xepapadeas, A. 1992. "Environmental Policy Design and Dynamic Nonpoint-Source Pollution." *Journal of Environmental Economics and Management*, 23, pp. 22–39.

2. Regulation of Pollution with Asymmetric Information

JEAN-JACQUES LAFFONT

GREMAQ and IDEI, Université de Toulouse, France

1. INTRODUCTION

It is now recognized that the regulation of externalities in an economy with a benevolent social regulator is a standard mechanism design problem which derives its difficulties from the various asymmetries of information existing between the regulator and the economic agents. Given the close analogy between a public good and an externality we can consider that this recognition goes back to Groves (1973). Dasgupta, Hammond and Maskin (1980), Laffont (1977) Ch. 7 were more explicit about the use of the Groves mechanisms to regulate externalities. Recently, the renewal of interest for environmental issues has further articulated the paradigm (Baron (1985a,b), Besanko (1987), Spulber (1989), Van Egteren (1989), Bowes, Macauley, Palmer (1991), Lewis-Sappington (1992), Segerson (1988), Xepapadeas (1991)). The purpose of this paper is to provide a general framework which encompasses a number of static issues in the regulation of externalities.

We start with the regulation of a natural monopoly which creates a negative externality. First we assume that the relationship between the level of production and the level of the externality can be modified by a costly technology and that the level of the externality is observable. This general case has been rarely considered in the literature which has been mostly concerned with situations in which there exists a fixed relationship between the level of production and the level of externality.

Having both the quantity of product and the pollution level observable makes the regulation problem similar to that of multiproduct regulation with one private

C. Dosi & T. Tomasi (eds.), Nonpoint Source Pollution Regulation: Issues and Analysis, 39–66.

good and one public good. We can then draw on Laffont-Tirole (1990) to obtain the optimal regulation under incomplete information.

The Pigouvian taxes must be modified in two ways. First, if the level of pollution affects the rate at which the firm can substitute the effort variable and the efficiency variable in its technology, then the level of pollution affects the size of the rent of asymmetric information captured by the firm. Consequently an incentive correction may appear to push the level of pollution in a direction which is favorable from the rent extraction viewpoint. Secondly, to mitigate the incentive problem, the regulator is led to reimburse a proportion, say a, of the cost to the firm. To preserve the right first order condition, the Pigouvian tax must then be scaled down by the factor $(1 - a)$.

Section 4 examines the difficulties created by the lack of observability of the pollution level and the need to base the tax on the related level of production.

Section 5 extends the analysis to nonpoint source pollution, both when the pollution comes from different plants of the same firm and when it comes from different firms. Conditions under which the lack of observability of personalized pollution levels creates no social cost are identified.

Finally, section 6 shows how various issues at the level of an industry add an additional public good dimension to the analysis and how the resolution of these public good problems interfere with optimal regulation.

2. THE BASIC MODEL

Consider a natural monopoly with the following cost function:

$$C = C(\beta, e, q, d) \qquad C_\beta > 0, C_e < 0, C_q > 0, C_d < 0 \tag{1}$$

where

q is the production level of the firm,

d is the level of pollution generated,

$\beta \in [\underline{\beta}, \bar{\beta}]$ is a productivity characteristic which is private information to the firm (adverse selection parameter),

e is the level of effort which decreases cost, but creates a disutility for the firm's manager $\psi(e)$ with $\psi\prime > 0, \psi\prime\prime > 0, \psi\prime\prime\prime \geq 0$ (moral hazard variable).

$\psi''' \geq 0$ ensures that stochastic regulation mechanisms are irrelevant. We assume that C is a strictly convex function of (q, d, e).[1]

Cost, price, quantity and pollution are observable by the regulator. Let $p(q)$ be the inverse demand function for the commodity produced by the monopoly and let $\eta(p)$ denote the price elasticity of demand. The consumers' welfare can be written:

$$W = S(q) - V(d) - p(q)q - (1 + \lambda)(C - p(q)q + t) \tag{2}$$

where

$S(q)$ is a strictly concave function describing the utility derived from q,

$V(d)$ is a strictly convex function describing the disutility due to pollution,

$1 + \lambda$ is the social price of public funds (due to distortionary taxation),

t is the transfer from the regulator to the firm,

and where we make the accounting convention that the regulator pays cost C directly and receives the proceeds from sales so that he must raise $C - p(q)q + t$ through distortionary taxation.

Because of our accounting convention the firm's welfare is

$$U = t - \psi(e) \tag{3}$$

and social welfare for a utilitarian regulator is

$$W + U = S(q) - V(d) + \lambda p(q)q - (1 + \lambda)(C + \psi) - \lambda U. \tag{4}$$

It is important to note that because of the social cost of raising public funds it is socially costly to give up a positive rent U to the monopoly.[2]

Under complete information (knowledge of β and observation of e) the regulator would maximize social welfare under the single participation constraint of the monopoly that we can normalize to

$$U \geq 0. \tag{5}$$

We assume that social welfare is strictly concave in (q, d, e). This is always obtained under our assumptions for λ small enough and enough concavity in the functions S and $-V$.

We obtain immediately

Proposition 1 *Under complete information optimal regulation is characterized by*

$$i) \quad \frac{p - C_q}{p} = \frac{\lambda}{1+\lambda} \frac{1}{\eta(p)}$$

$$ii) \quad \psi\prime(e) = -C_e$$

$$iii) \quad V\prime(d) = -(1+\lambda)C_d$$

$$iv) \quad U = 0.$$

Optimal regulation entails Ramsey pricing of the commodity,[3] an effort level which equates the marginal disutility of effort to its marginal cost improvement, a level of pollution which equates the marginal disutility of pollution to the marginal social cost of depollution, and no rent for the monopoly.

If p^*, d^*, q^*, e^* denote the levels of those variables for optimal regulation, an implementation of the optimal allocation is achieved by the following combination of instruments (if direct quantitative regulation is not desired):

A subsidy per unit of sold commodity of

$$\tau_1^* = \frac{p^*}{(1+\lambda)\eta(p^*)}, \tag{6}$$

a Pigouvian unit tax of the externality of

$$\tau_2^* = \frac{V\prime(d^*)}{1+\lambda}, \tag{7}$$

a lump sum tax on profit of

$$T^* = (p(q^*) + \tau_1^*)q^* - C(\beta, e^*, q^*, d^*) - \tau_2^* d^* - \psi(e^*). \tag{8}$$

The subsidy solves the monopolist's tendency to underproduce ; the Pigouvian tax leads the monopolist to internalize the costs of pollution correctly ; the lump sum tax solves the distributional problem.

Note that a lump sum tax is needed in order to elicit the optimal level of effort. The alternative of using a close to 100 % profit tax would not work here because the "cost" $\psi(e)$ is not monetary. Such a procedure would lead to an effort level that is much too low as the firm would maximize

$$\varepsilon\Big[(p(q)+\tau_1^*)q-\tau_2^*d-C(\beta,e,q,d)\Big]-\psi(e)$$

yielding $\psi\prime(e)=\varepsilon C_e$.

An alternative is to announce that an ε gain will be left to the firm if it follows the recommendations of the regulator, i.e., produces q^* and d^* with an effort level e^*.

3. REGULATION UNDER INCOMPLETE INFORMATION

Let us now consider the case of incomplete information when the regulator does not observe the effort level and does not know β. He has a prior distribution $F(.)$ over the support $[\underline{\beta},\bar{\beta}]$ with a strictly positive density function which satisfies the classical monotone hazard rate assumption $\dfrac{d}{d\beta}\left(\dfrac{F(\beta)}{f(\beta)}\right)\geq 0$.

Let $e=E(\beta,q,d,C)$ the solution in e of equation (1) defining the cost function. Substituting this expression in (3) we have

$$U=t-\psi\Big(E(\beta,q,d,C)\Big).$$

The problem then appears as an adverse selection problem with a parameter of private information β and three action variables q,d and C. From the revelation principle any regulation mechanism is equivalent to a revelation mechanism $t(\tilde{\beta}),q(\tilde{\beta}),d(\tilde{\beta}),C(\tilde{\beta})$ which specifies for any announcement $\tilde{\beta}\in[\underline{\beta},\bar{\beta}]$ a transfer from the regulator to the firm, a production level, a pollution level and a cost level to be realized, and which induces truthful revelation of the private information.

Let $U(\beta,\tilde{\beta})$ the utility level of a firm with characteristic β when it announces $\tilde{\beta}$. Then

$$U(\beta,\tilde{\beta})=t(\tilde{\beta})-\psi\Big(E(\beta,q(\tilde{\beta}),d(\tilde{\beta}),C(\tilde{\beta}))\Big). \tag{9}$$

The first order condition of incentive compatibility is

$$\frac{dt}{d\beta}(\beta)-\psi\prime\Big(E(\beta,q(\beta),d(\beta),C(\beta))\Big)\left(\frac{\partial E}{\partial q}\cdot\frac{dq}{d\beta}+\frac{\partial E}{\partial d}\cdot\frac{dd}{d\beta}+\frac{\partial E}{\partial C}\cdot\frac{dC}{d\beta}\right)=0 \tag{10}$$

and the local second order condition is:

$$-\psi'' E_\beta\left(\frac{\partial E}{\partial q}\cdot\frac{dq}{d\beta}+\frac{\partial E}{\partial d}\cdot\frac{dd}{d\beta}+\frac{\partial E}{\partial C}\cdot\frac{dC}{d\beta}\right)$$

$$-\psi'\left(\frac{\partial^2 E}{\partial q\partial\beta}\cdot\frac{dq}{d\beta}+\frac{\partial^2 E}{\partial d\partial\beta}\cdot\frac{dd}{d\beta}+\frac{\partial^2 E}{\partial C\partial\beta}\cdot\frac{dC}{d\beta}\right)\geq 0. \tag{11}$$

From our assumptions on the cost function $E_\beta > 0$, $\frac{\partial E}{\partial q} > 0, \frac{\partial E}{\partial d} < 0$, $\frac{\partial E}{\partial C} < 0$. If furthermore we assume $\frac{\partial^2 E}{\partial q\partial\beta} > 0, \frac{\partial^2 E}{\partial d\partial\beta} < 0$, and $\frac{\partial^2 E}{\partial C\partial\beta} < 0$, sufficient second order conditions are (see Guesnerie-Laffont (1984))

$$\frac{dq}{d\beta}\leq 0 \qquad \frac{dd}{d\beta}\geq 0 \qquad \frac{dC}{d\beta}\geq 0. \tag{12}$$

In a first step we will ignore these second order conditions but check *ex post* that they are satisfied under our assumptions by the solution we identify (see Appendix 1 for an example).

Let $U(\beta) = t(\beta) - \psi\Big(E(\beta, q(\beta), d(\beta), C(\beta))\Big)$ denote the rent of asymmetric information captured by firm β when it is truth-telling. The first order condition of incentive compatibility can then be rewritten in the simpler form

$$\dot{U}(\beta) = -\psi'(e(\beta))\;\; E_\beta(\beta, q(\beta), d(\beta), C(\beta)). \tag{13}$$

The participation constraint of the firm $U(\beta)\geq 0$ for any β can be reduced, in view of (13) which shows that the rent is decreasing in β, to

$$U(\bar\beta)\geq 0. \tag{14}$$

The regulator wishes to maximize under the incentive and participation constraints (13) and (14), expected social welfare, i.e.:

$$\int_{\underline\beta}^{\bar\beta}\Big[S(q(\beta)) - V(d(\beta)) + \lambda p(q(\beta))q(\beta) - (1+\lambda)\Big(C(\beta, e(\beta), q(\beta), d(\beta))$$

$$+\psi(e(\beta))\Big) - \lambda U(\beta)\Big]dF(\beta).$$

Solving this program yields the optimal regulation under incomplete information.

Proposition 2 *Under incomplete information optimal regulation is characterized by*

$$i) \quad \frac{p - C_q}{p} = \frac{\lambda}{1+\lambda}\frac{1}{\eta(p)} + \frac{\lambda}{1+\lambda}\frac{F(\beta)}{f(\beta)}\frac{\psi\prime}{p}.\frac{d}{dq}\Big(- \frac{C_\beta}{C_e}\Big)$$

$$ii) \quad \psi\prime(e) = -C_e - \frac{\lambda}{1+\lambda}\frac{F(\beta)}{f(\beta)}\frac{d}{de}(\psi\prime(e)E_\beta)$$

$$iii) \quad V\prime(d) = -(1+\lambda)C_d - \lambda\frac{F(\beta)}{f(\beta)}\psi\prime.\frac{d}{dd}\Big(- \frac{C_\beta}{C_e}\Big)$$

$$iv) \quad U(\beta) = \int_{\underline{\beta}}^{\bar{\beta}} \psi\prime(e(\tilde{\beta}))E_\beta\Big(\tilde{\beta}, q(\tilde{\beta}), d(\tilde{\beta}), C(\tilde{\beta})\Big)d\tilde{\beta}.$$

Proof: see Appendix 2.

The main features of the optimal allocation are as follows. The implicit optimal pricing rule deviates from the full information Ramsey pricing rule in two ways. First, the marginal cost C_q used in the definition of the Lerner index is taken at distorted values of effort and production levels. Secondly, there is an incentive correction which vanishes for the most efficient firm ($F(\underline{\beta}) = 0$) or if $\frac{d}{dq}\Big(- \frac{C_\beta}{C_e}\Big) = 0$. By an application of the aggregation Leontief theorem (see Laffont-Tirole (1990)), we know that $\frac{d}{dq}(-C_\beta/C_e) = 0$ is equivalent to the existence of a function $\xi(.)$ such that $C = C(\xi(\beta, e, d), q, d)$. In general the pricing rule participates in the extraction of the firm's rent of asymmetric information, but the incentive correction vanishes when the production level (or the price level) does not affect $E_\beta = -C_\beta/C_e$, which defines the rate at which the firm can substitute β and e, i.e. the manipulation of its efficiency characteristic and effort, for given levels of C, q and d. Then, we say that the dichotomy property holds (see Laffont-Tirole (1990)). As shown by (13) E_β plays a crucial role in the definition of the rent. There is no point in distorting the price for incentive reasons if the quantity does not affect the rent of asymmetric information.

Equation ii) in Proposition 2 can be interpreted as a cost reimbursement rule $t(C)$. Then the firm maximizes with respect to e

$$t(C(\beta, e, q, d)) - \psi(e)$$

yielding $\frac{dt}{dC}.C_e = \psi\prime(e)$.

The marginal share of cost supported by firm β in the net transfer is

$$\frac{dt}{dC} = -1 - \frac{\lambda}{1+\lambda}.\frac{F(\beta)}{f(\beta)}.\frac{1}{C_e}.\frac{d}{de}(\psi\prime(e)E_\beta)$$

which actually means that a proportion

$$\frac{\lambda}{1+\lambda}\frac{F}{f}\frac{1}{(-C_e)}\frac{d}{de}(\psi\prime(e)E_\beta)$$

is reimbursed to the firm. Consequently, except for the most efficient type, the level of effort is lower than under complete information (for given q and d). By decreasing effort, $\psi\prime E_\beta$ is decreased and therefore the rent is decreased (see iv in Proposition 2).

The allocation rule for pollution is altered when $\frac{d}{dd}E_\beta$ does not vanish. The incentive correction vanishes if a function ξ exists such that $C = C(\xi(\beta, e, q), d, q)$. Then changing the level of pollution does not affect the rent of asymmetric information.

Incentives affect neither the pricing rule nor the pollution rule if, and only if, there exists a function Φ such that $C = C(\Phi(\beta, e), d, q)$. Then, a dichotomy exists between the cost reimbursement rule defined by ii) which optimizes the extraction of the rent and the production and pollution rules: these latter follow the same principles as under complete information, but with different levels for the variables due to the distortions in effort induced by the cost reimbursement rule.

The sign of the incentive correction for the pollution rule is determined by the sign of $\frac{d}{dd}E_\beta$. If increasing the level of pollution d increases E_β (and therefore the rent), the incentive correction is positive.

Finally, iv) tells us that optimal regulation gives up a rent of asymmetric information to the monopoly which decreases with β. The optimal regulation trades-off optimally the cost due to the rent of asymmetric information and the allocative distortions of q, d and e.

Let us now come to a more concrete implementation of optimal regulation.

One possibility is to announce to the firm a cost reimbursement rule $T(C)$ complemented by production and pollution rules $q(C)$ and $d(C)$. The firm then maximizes

$$T(C) - \psi(E(\beta, q(C), d(C), C))$$

and picks C such that

$$\frac{dT}{dC} - \psi\prime\left(E_q\frac{dq}{dC} + E_d\frac{dd}{dC} + E_C\right) = 0$$

which is equivalent to the first order condition of incentive compatibility

$$\frac{dT}{d\beta} - \psi\prime\left(E_q\frac{dq}{d\beta} + E_d\frac{dd}{d\beta} + E_C\frac{dC}{d\beta}\right) = 0$$

if the function $T(C)$ is chosen as $t(\beta^{-1}(C))$ with $\beta^{-1}(C)$ being the inverse function of $C(\beta)$, and if

$$q(C) = q(\beta^{-1}(C)),$$

and

$$d(C) = d(\beta^{-1}(C)).$$

A simpler implementation is obtained when the dichotomy property holds for example for the special cost function

$$C = (\beta - e)G(q, d) \tag{15}$$

with $G_q > 0,\ G_d < 0$ and G convex in (q, d).

From (15), $e = \beta - \dfrac{C}{G(q,d)} \equiv \beta - H(C, q, d)$.

The firm's objective function is then

$$t - \psi(\beta - H(C, q, d)).$$

What matters for incentive purposes is only the revelation mechanism $t(\tilde{\beta}), H(\tilde{\beta})$ with first order condition

$$\frac{dt}{d\beta} + \psi\prime\frac{dH}{d\beta} = 0$$

and second order condition

$$\frac{dH}{d\beta} \geq 0.$$

When H is increasing the optimal revelation mechanism $(t(\beta), H(\beta))$ can be replaced by the non linear transfer

$$T(H) = t(\beta^{-1}(H)).$$

The optimal regulation is here achieved by giving to the firm a non linear transfer $T\left(\frac{C}{G(q,d)}\right)$ which is a function of the cost level achieved, the production level and the pollution level. Then the firm is recommended to choose particular production, pollution and effort levels. It is indifferent to those levels which produce the same aggregate $C/G(q,d)$. As here $T(.)$ can easily be shown to be convex (Laffont-Tirole, 1990), the non-linear function $T(.)$ can be replaced by a menu of linear transfer rules of the type

$$T(H,H^a) = A(H^a) - B(H^a)(H - H^a)$$

where

$$H^a \equiv \frac{C(\beta^a)}{G(q(\beta^a), d(\beta^a))}$$

is the "announced" level of **performance** and H the achieved one.

Finally, note that if a zero mean random error or shock $\tilde{\varepsilon}$ with a distribution independent of β and e is introduced in the cost function, $C = (\beta - e)G(q,d) + \tilde{\varepsilon}$, risk neutrality implies that the same allocation can be achieved as before because of implementation by a menu of **linear** transfers.

Let us now describe an implementation in terms of the more traditional tax instruments. The firm is asked to announce a cost target C and is promised

- a production subsidy $\tau_1^*(C)$
- a pollution tax $\tau_2^*(C)$
- a transfer $\hat{t}(C)$, if it produces C.

It then maximizes with respect to C, q and d

$$\hat{t}(C) + (p(q) + \tau_1^*(C))q - \tau_2^*(C)d - \psi(E(\beta, q, d, C)).$$

Denoting t the global transfer $\hat{t} + \tau_1^* q - \tau_2^* d$ we obtain in particular:

$$\frac{dt}{dC} = \psi\prime E_C$$

$$\tau_2^*(C) = -\psi\prime E_d.$$

Requiring

$$\frac{dt}{dC} = -1 + \frac{\lambda}{1+\lambda}\frac{F}{f}\frac{1}{(-C_e)}.\frac{d}{de}(\psi\prime(e)E_\beta)$$

and since

$$\psi\prime E_d = \frac{dt}{dC}\frac{E_d}{E_C} = \frac{dt}{dC}.C_d$$

$$\tau_2^*(C) = -\frac{dt}{dC}.C_d = \frac{dt}{dC}.\left[\frac{V\prime(d)}{1+\lambda} + \frac{\lambda}{1+\lambda}\frac{F}{f}\psi\prime\frac{d}{dd}E_\beta\right].$$

The Pigouvian tax is now modified in several ways. As explained above there is an incentive correction which has the sign of $\frac{d}{dd}E_\beta$. Moreover, as the firm β is now supporting only a share $\frac{dt}{dC}$ of its cost, the Pigouvian tax is proportionally decreased by the same factor.

The intuition of optimal regulation under incomplete information is now clear.

To mitigate the rent of asymmetric information, the regulator reimburses a share of the cost to the firm which increases with β. This leads to a lower effort level for less efficient firms and, as this course of action is less tempting for more efficient firms, it is possible to give smaller rents for the latter.

The appropriate production and pollution levels can be decentralized by a production subsidy and a pollution tax, which must be appropriately scaled down given that the firm bears only a fraction of its cost. In addition to this scaling factor, incentive corrections are needed if varying production or pollution levels affect the rate at which the firm can substitute effort and efficiency. In the case of several firms, the classical equivalence between taxes and permits to pollute could be recovered as follows: in the case of taxes directly controlled by the regulator the scaling of taxes can be personalized by the regulator. In the case of a market for pollution permits the required personalized scaling of pollution prices would need personalized subsidies for the transactions taking place on the market for pollution permits.

4. REGULATION WITH NON VERIFIABLE POLLUTION LEVEL

So far we have generalized the notion of Pigouvian tax to take into account asymmetric information. There the tax was based on the observable level of pollution. At an early stage it was recognized in the literature (Plott (1966)), that the appropriate variable to tax is often not observable, and that taxing inappropriate variables is sub-optimal.

Suppose here that the technology is known to the regulator (up to β and e) but that the level of pollution d is not verifiable. The variable d plays the role of an additional moral hazard variable.

The objective function of the firm is now

$$t - \psi(E(\beta, C, q, d)).$$

Clearly the firm's aim is to maximize pollution to decrease its cost, and the regulator has no instruments to avoid this.

There are several channels by which regulation can nevertheless affect pollution.

Suppose that the maximal amount of pollution is a function of the production level $\bar{d}(q)$. The firm's aim is to produce this maximum level of pollution, but the optimal regulatory scheme will now be the solution of

$$\text{Max} \int_{\underline{\beta}}^{\bar{\beta}} \Big[S(q(\beta)) - V(\bar{d}(q(\beta)) + \lambda p(q(\beta))q(\beta)$$

$$-(1+\lambda)\Big(C(\beta, e(\beta), q(\beta), \bar{d}(q(\beta))) + \psi(e(\beta))\Big) - \lambda U(\beta)\Big] dF(\beta)$$

s.t.

$$\dot{U} = -\psi\prime(e(\beta))E_\beta\Big(\beta, q(\beta), \bar{d}(q(\beta)), C(\beta)\Big)$$

$$U(\bar{\beta}) \geq 0.$$

The pricing equation is now

$$\frac{p - C_q}{p} = \frac{\lambda}{1+\lambda}\frac{1}{\eta(p)} + \frac{\lambda}{1+\lambda}\frac{F(\beta)}{f(\beta)}\frac{\psi\prime}{p}\frac{d}{dq}E_\beta$$

$$+\left(\frac{V\prime}{1+\lambda} + C_d + \frac{\lambda}{1+\lambda}\frac{F(\beta)}{f(\beta)}\psi\prime\frac{d}{dd}E_\beta\right)\frac{\bar{d}\prime}{p}.$$

$$\tau_3 = \left(\frac{V\prime}{1+\lambda} + C_d + \frac{\lambda}{1+\lambda}\frac{F}{f}\psi\prime\frac{d}{dd}E_\beta\right)\bar{d}\prime$$

can be interpreted (up to the appropriate scaling factor) as a "tax" on the good produced by the monopoly to indirectly decrease pollution. Note from iii) in Proposition 2 that this tax is zero if the level of pollution is optimized separately.

This tax equals the social marginal disutility of the pollution created by the marginal unit of production which includes

$\dfrac{V'\bar{d}'}{1+\lambda}$: the marginal direct disutility of pollution

$C_d\bar{d}'$: the marginal "gain" in cost

$\dfrac{\lambda}{1+\lambda}\dfrac{F}{f}\psi'\dfrac{d}{dd}E_\beta\bar{d}'$: the marginal disutility or gain (depending on the sign of $\dfrac{d}{dd}E_\beta$) due to the rent effect.

A byproduct of this analysis is the optimal regulation when the level of pollution is strictly related to the production level by the relation $d(q)$. We then obtain the optimal tax on the commodity produced which defines the optimal regulation by replacing $\bar{d}(.)$ by $d(.)$. In that case a single instrument is enough. A remark analogous to Buchanan (1969) who observed that the underproduction of a monopoly and the over production due to an externality may compensate each other, is here that optimal pricing can reduce to Ramsey pricing if the incentive distortion and the externality tax compensate each other.

Suppose now that the marginal utility of the commodity is affected by pollution, and let us consider a utility function $S(q,d)$. If the price level is observable (as well as the quantity) the pollution level can be easily inferred from

$$p = \frac{\partial S}{\partial q}(q,d) \equiv p(q,d)$$

and consequently controlled as if pollution was observable.

In practice, the regulator does not know the utility function as well as the firm and an additional asymmetry of information makes the observation of price an imperfect control of pollution (see Laffont-Tirole (1991a) where this approach is followed to model quality).

Let us here make the extreme assumption that price is not verifiable so that the indirect observation of pollution cannot be used. It is no longer legitimate to reason in net transfers, since sales are not observable. The objective function of the firm must be written:

$$t(\beta) + p(q(\beta),d)q(\beta) - \psi\Big(E(\beta, C(\beta), q(\beta), d)\Big).$$

The incentive constraint remains

$$\dot{U} = -\psi' E_\beta$$

but we must now add the moral hazard constraint

$$q(\beta)\frac{\partial p}{\partial d}\,(q(\beta),d) - \psi\prime E_d = 0.$$

The regulator's optimization program is now

$$\text{Max} \int_{\underline{\beta}}^{\bar{\beta}} \Big[S(q(\beta),d(\beta)) + \lambda p(q(\beta),d(\beta))q(\beta)$$

$$-(1+\lambda)\Big(C(\beta,e(\beta),q(\beta),d(\beta)) + \psi(e(\beta))\Big) - \lambda U(\beta)\Big]dF(\beta)$$

$$\dot{U}(\beta) = -\psi\prime(e(\beta)).E_\beta\Big(\beta, C(\beta,e(\beta),q(\beta),d(\beta)),q(\beta),d(\beta)\Big)$$

$$U(\bar{\beta}) \geq 0$$

$$q(\beta)\frac{\partial p}{\partial d}(q(\beta),d(\beta)) - \psi\prime(e(\beta))E_d\Big(\beta, C(\beta,e(\beta),q(\beta),d(\beta)),q(\beta),d(\beta)\Big) = 0.$$

If $\mu(\beta)$ is the multiplier associated with the state variable U, from the Pontryagin principle $\mu(\beta) = \lambda F(\beta)$. Let $\nu(\beta)$ be the multiplier associated with the moral hazard constraint. First order conditions of optimal regulation are:

$$\left(\frac{\partial S}{\partial q} + \lambda\left(\frac{\partial p}{\partial q}.q + p\right) - (1+\lambda)C_q\right)f$$

$$-\lambda F\psi\prime\frac{d}{dq}E_\beta + \nu\left(\frac{\partial p}{\partial d} + q\frac{\partial^2 p}{\partial d\partial q} - \psi\prime\frac{d}{dq}E_d\right) = 0 \qquad (16)$$

$$\left(\frac{\partial S}{\partial d} + \lambda\frac{\partial p}{\partial d}.q - (1+\lambda)C_d\right)f$$

$$-\lambda F\psi\prime\frac{d}{dd}E_\beta + \nu\left(q\frac{\partial^2 p}{\partial d^2} - \psi\prime\frac{d}{dd}E_d\right) = 0 \qquad (17)$$

$$-(1+\lambda)(\psi\prime + C_e)f - \lambda F\frac{d}{de}(\psi\prime E_\beta) - \nu\frac{d}{de}(\psi\prime E_d) = 0. \qquad (18)$$

Note that little can be said in general about this solution in comparison with the case of full control. Depending on the comparison between the level of pollution achieved here and the one obtained with direct control of pollution, from the first order condition in d and the second order condition of the firm's problem, we derive the sign of the multiplier ν. Then we obtain the signs of the price and effort distortions.

For example, if pollution is too high (as it should be in most cases), the optimization defined by the above program is unchanged if the moral hazard constraint $q\frac{\partial p}{\partial d} - \psi\prime E_d = 0$ is replaced by $q\frac{\partial p}{\partial d} - \psi\prime E_d \leq 0$ (Indeed this new constraint only allows higher levels of pollution that are not socially desired). We can then conclude that the Kuhn and Tucker multiplier is non-positive. Alternatively, from (17) and from the second order condition in d of the firm's optimization program, $\nu \leq 0$.

If $\frac{d}{de}(\psi\prime(E_d)) > 0$ (respectively < 0) the non observability of pollution pushes towards more (respectively less) incentive for effort because a higher level of effort makes a marginal decrease of pollution less costly to the firm.

Example: If

$$C(\beta, e, q, d) = (\beta - e)(q - d) \ , \ E = \beta - \frac{C}{q - d}$$

$$E_d = -\frac{C}{(q-d)^2} \ , \ \frac{d}{de}E_d = -\frac{C_e}{(q-d)^2} > 0$$

$$\frac{d}{de}(\psi\prime e_d) = -\frac{C\psi\prime\prime}{(q-d)^2} - \frac{\psi\prime C_e}{(q-d)^2} = -\frac{\psi\prime C}{e(q-d)^2}\left[e\frac{\psi\prime\prime}{\psi\prime} - e\frac{(-C_e)}{C}\right]$$

If the effort elasticity of the marginal disutility of effort ($\psi\prime$) is higher than the effort elasticity of cost, incentives are enhanced.

Remark: If cost is unobservable, no sharing of cost is possible and the firm fully internalizes the effect of its effort choice. The appropriate revelation mechanism is then $t(\tilde{\beta}), q(\tilde{\beta}), d(\tilde{\beta})$ and the only way to mitigate the rent of asymmetric information is to distort the pricing rule or the pollution rule. In that case there is no need to scale down the Pigouvian taxes. The optimal mechanism can be implemented by a non linear transfer $t(q)$ and a non linear pollution rule $d(q)$ or a non linear transfer $t(d)$ and a non linear production rule $q(d)$ (See Appendix 3 for additional information).

5. NONPOINT SOURCE POLLUTION

5.1. Several Plants Belonging to the Same Firm

Suppose now that two plants belonging to the same firm contribute to pollution with cost technologies

$$C_1 = C_1(\beta, e_1, q_1, d_1) \tag{19}$$

$$C_2 = C_2(\beta, e_2, q_2, d_2) \tag{20}$$

where we assume that a single parameter β describes the firm's private information, e_i is the level of effort exerted in plant i, q_i is the production level of plant i and d_i its pollution level.

Suppose first that d_1 and d_2 are observable.

The firm's utility level is

$$U = t - \psi(e_1 + e_2).$$

Social welfare is

$$\begin{gathered} S(q_1 + q_2) - V(d_1 + d_2) + \lambda p(q_1 + q_2)(q_1 + q_2) \\ -(1 + \lambda)\Big(C_1(\beta, e_1, q_1, d_1) + C_2(\beta, e_2, q_2, d_2) + \psi(e_1 + e_2)\Big) \\ -\lambda U. \end{gathered}$$

Denoting $E_1(\beta, C_1, q_1, d_1)$ and $E_2(\beta, C_2, q_2, d_2)$ the solutions in e_1 and e_2 of (19) (20), the first order incentive constraint is:

$$\begin{aligned} \dot{U} &= -\psi\prime(e_1 + e_2)\frac{\partial}{\partial\beta}(E_1 + E_2) \\ &\equiv -\psi\prime(e_1 + e_2)(E_{1\beta} + E_{2\beta}). \end{aligned}$$

Maximizing expected social welfare under incentive and participation constraints gives in particular

$$\frac{V\prime(d_1 + d_2)}{1 + \lambda} + \frac{\lambda}{1 + \lambda}\,\frac{F(\beta)}{f(\beta)}\,\psi\prime\frac{d}{dd_1}E_{1\beta} = -C_{d_1} \tag{21}$$

$$\frac{V\prime(d_1 + d_2)}{1 + \lambda} + \frac{\lambda}{1 + \lambda}\,\frac{F(\beta)}{f(\beta)}\,\psi\prime\frac{d}{dd_2}E_{2\beta} = -C_{d_2} \tag{22}$$

(21) and (22) can be implemented by per unit taxes on pollution proportional to

$$\tau_1 = \frac{V\prime(d_1 + d_2)}{1+\lambda} + \frac{\lambda}{1+\lambda}\frac{F}{f}\psi\prime\frac{d}{dd_1}E_{1\beta}$$

$$\tau_2 = \frac{V\prime(d_1 + d_2)}{1+\lambda} + \frac{\lambda}{1+\lambda}\frac{F}{f}\psi\prime\frac{d}{dd_2}E_{2\beta}.$$

In general it is useful to observe the individual pollution levels because the terms $\frac{d}{dd_1}E_{1\beta}$ and $\frac{d}{dd_2}E_{2\beta}$ are different. As pollution levels d_1 and d_2 may have different effects on the rates at which efforts e_1 and e_2 must be substituted for manipulations of β, they affect differently the rent which must be given to the firm. The optimal trade-off between rent and allocative inefficiencies call for different taxes for pollutions d_1 and d_2.

From (21) (22) we see that the optimal taxes are equal, and therefore observability of individual pollution levels is irrelevant if a constant K exists such that

$$K = \frac{d}{dd_1}E_{1\beta} = \frac{d}{dd_2}E_{2\beta}.$$

When $K = 0$, the characterization of cost functions satisfying this property is easily achieved by observing that $E_{1\beta} = \frac{-C_{1\beta}}{C_{1e_1}}$ and $E_{2\beta} = \frac{-C_{2\beta}}{C_{2e_2}}$ and by using Leontief's aggregation theorem. Functions ξ_1 and ξ_2 must exist so that

$$C_1 = C_1\Big(\xi_1(\beta, e, q_1), q_1, d_1\Big)$$

$$C_2 = C_2\Big(\xi_2(\beta, e, q_2), q_2, d_2\Big)$$

The uniform tax which decentralizes the optimal pollution levels is (up to the appropriate scaling factor):

$$\tau = \frac{V\prime(d_1 + d_2)}{1+\lambda}.$$

5.2. *Several Plants Belonging to Different Firms*

Consider now the case of two different firms producing two different goods (yielding consumers' gross surpluses $S_1(q_1), S_2(q_2)$) with technologies

$$C_1 = C_1(\beta_1, e_1, q_1, d_1)$$

$$C_2 = C_2(\beta_2, e_2, q_2, d_2)$$

where β_1 and β_2 are independently drawn from distributions $F_1(.)$ and $F_2(.)$.

When d_1 and d_2 are observable, optimal regulation is simply the juxtaposition of two regulations as defined in Section 2.

We will find conditions on the technologies so that optimal regulation can be achieved despite the non observability of individual pollution levels, i.e., the case of nonpoint pollution where only $d = d_1 + d_2$ is observable.

Consider the particular technologies which allow implementation of the optimal regulation by menus of linear contracts (see Laffont-Tirole (1990) and Section 2)

$$C_1 = (\beta_1 - e_1)G_1(q_1, d_1)$$

$$C_2 = (\beta_2 - e_2)G_2(q_2, d_2)$$

Regulation of firm i when d_i, C_i, q_i are observable is implemented through the non linear tax $T(\frac{C_i}{G_i(q_i,d_i)})$. As $T(.)$ is here convex, it can be replaced by a menu of linear transfers

$$T(H_i, H_i^a) = A(H_i^a) - B(H_i^a)(H_i - H_i^a)$$

with

$$H_i^a = \frac{C(\beta_i^a)}{G(q_i(\beta_i^a), d_i(\beta_i^a))}.$$

Consider further the particular case where

$$H_i^a = \frac{C(\beta_i^a)}{\tilde{G}(q_i(\beta_i^a))} d_i(\beta_i^a)$$

i.e. H_i^a is linear in d_i.

Now, if d_i is not observed but only $d = d_1 + d_2$ is observed,[4] the following transfer function still implements the same allocation.

Let $\bar{d}_j = E_{\beta j} d_j(\beta_j) \quad j \neq i$ and

$$T(H_i^a, C_i, q_i, d) = A(H_i^a) - B(H_i^a)\Big[\frac{C_i}{\tilde{G}(q_i)}(d - \bar{d}_j) - H_i^a\Big].$$

As the firm is risk neutral,

$$E_{\beta_j}T(H_i^a, C_i, q_i, d) = A(H_i^a) - B(H_i^a)\Big[\frac{C_i d_i}{\tilde{G}(q_i)} - H_i^a\Big]$$
$$= A(H_i^a) - B(H_i^a)(H_i - H_i^a)$$
$$= T(H_i, H_i^a).$$

For cost functions of the type

$$C_i = (\beta_i - e_i)\frac{\tilde{G}(q_i)}{d_i}$$

nothing is lost from the lack of individual observability due to nonpoint source pollution, as long as the firms are risk neutral and are not subject to an ex post liability constraint. Clearly, those assumptions can be approximately valid only in cases with a small number of polluters. Of course, a good reason must exist to justify the inability to observe individual pollution levels in this small number situation. It can be technological or political. Agency problems in the observation of individual pollution levels may justify the use of mechanisms based on global pollution. The approach of Laffont-Tirole (1991b) could be used to model this idea.

6. REGULATION OF POLLUTION IN AN INDUSTRY

The optimization of pollution activities in an industry has often two dimensions: firstly each firm must generate an appropriate level of pollution, secondly a collective decision of a public good nature must be taken. As an example the number of firms must be induced optimally through appropriate entry incentives in the industry (see Schultze-D'Arge (1974), Carlton-Loury (1980), Spulber (1985)). Another example is the problem of the localization of firms when the spatial distribution of pollution matters (Braden, Bouzaher, Johnson and Miltz (1989), Graham-Tomasi (1992)).

In this section we give an example of an analysis which combines the optimization of pollution levels and localization of firms.

For simplicity, we consider the case of two firms as in section 5.2. Firm 1 must locate at point a and firm 2 can locate a proportion x_2^b of its activity at point b (the complement at point a). It then suffers an additional cost Δx_2^b.

Consequently pollution at points a and b is

$$d^a = d_1 + d_2(1 - x_2^b)$$

$$d^b = d_2 x_2^b.$$

Let $V(d_1 + d_2, |d^a - d^b|)$ be the disutility of pollution. It depends on the total level of pollution, $d_1 + d_2$, and on the location of pollution. V is increasing in the second argument. It is better if the pollution is spread over the two locations equally. As we assume that it is more costly for firms to locate in b than in a so that it will be always socially better to have $d^a \geq d^b$, we can write the disutility of pollution

$$V(d_1 + d_2, d^a - d^b)$$

or

$$V(d_1 + d_2, d_1 + d_2(1 - 2x_2^b)).$$

Under complete information, optimal regulation is defined by:

$$\text{Max}\Big\{S(q_1 + q_2) - V(d_1 + d_2, d_1 + d_2(1 - 2x_2^b)) + \lambda p(q_1 + q_2)(q_1 + q_2)$$
$$-(1 + \lambda)\Big(C_1(\beta_1, e_1, q_1, d_1) + C_2(\beta_2, e_2, q_2, d_2) + \psi(e_1) + \psi(e_2)$$
$$+\Delta x_2^b\Big)\Big\}$$

with first order conditions (assuming an interior maximum):

$$S\prime(q_1 + q_2) + \lambda(p + (q_1 + q_2)p\prime) = (1 + \lambda)C_{1q_1} = (1 + \lambda)C_{2q}$$

$$\psi\prime(e_1) = -C_{1e_1}$$

$$\psi\prime(e_2) = -C_{2e_2}$$

$$\frac{V_1' + V_2'}{1 + \lambda} = -C_{1d_1}$$

$$\frac{V_1' + (1 - 2x_2^b)V_2'}{1 + \lambda} = -C_{2d_2}$$

$$\frac{2d_2V_2'}{1+\lambda} = \Delta$$

where all variables are evaluated at the optimum $*$. The decentralization of this allocation is obtained by personalized Pigouvian taxes

$$\tau_{21} = \frac{V_1'^* + V_2'^*}{1+\lambda} \qquad \tau_{22} = \frac{V_1'^* + (1-2x_2^{b*})V_2'^*}{1+\lambda},$$

by subsidies per unit of commodity sold

$$\tau_1 = \frac{p^*}{(1+\lambda)\eta(p^*)},$$

a subsidy τ_3 for the share of activity located at b by firm 2

$$\tau_3 = \frac{2d_2^*V_2'^*}{1+\lambda}$$

and a lump sum tax which takes away all the profit.

Under incomplete information we assume that the distributions $F_1(\beta_1)$ and $F_2(\beta_1)$ of β_1 and β_2 on $[\underline{\beta}, \bar{\beta}]$ are independent and common knowledge and we postulate Bayesian Nash behavior for both firms. From the revelation principle we know that any regulation mechanism is equivalent to a revelation mechanism which specifies for each announcements (β_1, β_2)

- production levels $q_1(\beta_1,\beta_2), q_2(\beta_1,\beta_2)$,
- pollution levels $d_1(\beta_1,\beta_2), d_2(\beta_1,\beta_2)$,
- a proportion of firms 2's activity $x_2^b(\beta_1,\beta_2)$ located at b,
- compensatory transfers $t_1(\beta_1,\beta_2), t_2(\beta_1,\beta_2)$

(maintaining the convention of reasoning in net transfers).

Firm 1's rent is :

$$U_1(\beta_1,\beta_2) = t_1(\beta_1,\beta_2) - \psi(E_1(\beta_1, C_1, q_1, d_1))$$

Denoting $U_1(\beta_1) = E_{\beta_2}U_1(\beta_1,\beta_2)$, we obtain by the envelope theorem the incentive constraint

$$\dot{U}_1(\beta_1) = -E_{\beta_2}\psi\prime(e_1)\frac{\partial E_1}{\partial \beta_1} \tag{23}$$

with a similar expression for firm 2

$$\dot{U}_2(\beta_2) = -E_{\beta_1}\psi\prime(e_2)\frac{\partial E_2}{\partial \beta_2}. \tag{24}$$

Solving (20) (21) with the boundary conditions $U_1(\bar{\beta}) = U_2(\bar{\beta}) = 0$ and substituting in expected social welfare we obtain

$$\int_{\underline{\beta}}^{\bar{\beta}}\int_{\underline{\beta}}^{\bar{\beta}}\Big\{S(q_1(\beta_1,\beta_2)+q_2(\beta_1,\beta_2))+\lambda p(q_1(\beta_1,\beta_2)$$
$$+q_2(\beta_1,\beta_2))(q_1(\beta_1,\beta_2)+q_2(\beta_1,\beta_2))$$
$$-V\Big(d_1(\beta_1,\beta_2)+d_2(\beta_1,\beta_2), d_1(\beta_1,\beta_2)+d_2(\beta_1,\beta_2)(1-2x_2^b(\beta_1,\beta_2)\Big)$$
$$-(1+\lambda)\Big[(C_1(\beta_1,e_1(\beta_1,\beta_2),q_1(\beta_1,\beta_2),d_1(\beta_1,\beta_2))+\psi(e_1(\beta_1,\beta_2))$$
$$+C_2(\beta_2,e_2(\beta_1,\beta_2),q_2(\beta_1,\beta_2),d_2(\beta_1,\beta_2))+\psi(e_2(\beta_1,\beta_2))+\Delta x_2^b(\beta_1,\beta_2))\Big]$$
$$-\lambda\frac{F(\beta_1)}{f(\beta_1)}\psi\prime(e_1(\beta_1,\beta_2))\frac{\partial E_1}{\partial\beta_1}$$
$$-\lambda\frac{F(\beta_2)}{f(\beta_2)}\psi\prime(e_2(\beta_1,\beta_2))\,\frac{\partial E_2}{\partial\beta_2}\Big\}\;dF(\beta_1)dF(\beta_2)$$

yielding the first order conditions

$$S\prime(q_1+q_2)+\lambda(p\prime(q_1+q_2)(q_1+q_2)+p(q_1+q_2))$$
$$=(1+\lambda)C_{1q_1}+\lambda\frac{F(\beta_1)}{f(\beta_1)}\psi\prime(e_1)\frac{d}{dq_1}.\frac{\partial E_1}{\partial\beta_1}$$
$$=(1+\lambda)C_{2q_2}+\lambda\frac{F(\beta_2)}{f(\beta_2)}\psi\prime(e_2)\frac{d}{dq_2}.\frac{\partial E_2}{\partial\beta_2}$$

$$\frac{V_1'+V_2'}{1+\lambda}=-C_{1q_1}+\frac{\lambda}{1+\lambda}\frac{F(\beta_1)}{f(\beta_1)}\psi\prime(e_1)\frac{d}{dd_1}\frac{\partial E_1}{\partial\beta_1}$$

$$\frac{V_1'+(1-2x_2^b)V_2'}{1+\lambda}=-C_{2q_2}+\frac{\lambda}{1+\lambda}\frac{F(\beta_2)}{f(\beta_2)}\psi\prime(e_2)\frac{d}{dd_2}\frac{\partial E_2}{\partial\beta_2}$$

$$2d_2V_2'=\Delta(1+\lambda).$$

The implementation of this optimal regulation can be achieved as described in section 3 with non linear transfers for each firm.

Let us note simply that incomplete information does not affect the first order condition concerning location, because we have assumed that the location costs are independent of production cost. Therefore the choice of x_2^b does not affect incentives and there is no point in distorting x_2^b for incentive reasons.

As this section shows, there is no particular difficulty of introducing the additional public good dimensions of an optimal regulation of the industry in the analysis. The interaction between location subsidies (or taxes) and regulation is the same as the interaction with Pigouvian taxes described before. In so far as the firm is bearing only a fraction of its costs for incentive reasons the location subsidy must be also scaled down to yield the right first order conditions.

APPENDIX 1. AN EXAMPLE

$$V(d) = 2d^2$$

$$C = (\beta - e)(q - d)$$

$$p = 1 - q$$

$$\psi(e) = \frac{1}{2}e^2$$

For λ close enough to one, the regulator's program under complete information is concave.

The first order equations of optimal regulation under complete information are

$$1 - q - \beta + e = \frac{\lambda}{1+\lambda}(1-q)q$$

$$e = q - d$$

$$4d = (1+\lambda)(\beta - e).$$

Differentiating we see that for λ close to one,

$$\frac{dq}{d\beta} < 0 \quad , \quad \frac{dd}{d\beta} > 0 \quad , \quad \frac{de}{d\beta} < 0.$$

APPENDIX 2

The Hamiltonian of the program is:

$$H = \Big[S - V + \lambda p(q)q - (1+\lambda)(C+\psi) - \lambda U\Big] f - \mu\psi\prime E_{\beta}$$

where μ is co-state variable associated with U. From the Pontryagin principle

$$\dot{\mu}(\beta) = -\frac{\partial H}{\partial U} = \lambda f.$$

Using the transversality condition $\mu(\underline{\beta}) = 0$, we derive $\mu(\beta) = \lambda F(\beta)$.

Differentiating with respect to q, d and e yields immediately i) ii), iii) in Proposition 2. Integrating the incentive constraint (13) with the constraint $U(\bar{\beta}) = 0$ yields iv).

When $\lambda = 0$, we can find assumptions so that

$$\frac{dq}{d\beta} \leq 0, \ \frac{dd}{d\beta} \geq 0, \ \frac{de}{d\beta} \geq 0$$

(see Appendix 1 for an example). If λ is small enough, it remains so under incomplete information. The second order conditions of incentive compatibility are then satisfied.

APPENDIX 3. NON OBSERVABILITY OF COST

Since effort is optimal conditionally to the production and pollution levels (the firm minimizes $C + \psi$) we can as well suppress the effort variable and consider the technology $C = C(\beta, q, d)$ with $C_\beta > 0, C_q > 0, C_d < 0$. Then

$$U = t - C(\beta, q, d)$$

and the first order incentive compatibility condition is

$$\dot{U} = -C_\beta(\beta, q(\beta), d(\beta))$$

with the sufficient second order conditions

$$\frac{dq}{d\beta} \leq 0 \quad \frac{dd}{d\beta} \geq 0.$$

Maximizing the regulator's expected utility under the incentive and individual rationality constraints we obtain

$$\frac{p - Cq}{p} = \frac{\lambda}{1+\lambda}\frac{1}{\eta(q)} + \frac{\lambda}{1+\lambda}\frac{F}{f} C_{\beta q}$$

$$\frac{V'}{1+\lambda} = -C_d - \frac{\lambda}{1+\lambda}\frac{F}{f} C_{\beta d}.$$

The optimal mechanism can be implemented by a non linear transfer function $t(q)$ and a Pigouvian tax $\tau(q)d$ with

$$\tau(q) = \frac{V'}{1+\lambda} + \frac{\lambda}{1+\lambda}\frac{F}{f} C_{\beta d}.$$

To the usual marginal disutility of pollution $\frac{V'}{1+\lambda}$ is added a term to take into account the cost of the rent. If $C_{\beta d} > 0 (< 0)$ an increase of d increases (decreases) the rent (since $\dot{U} = -C_\beta$) ; therefore the pollution tax is increased (decreased) to decrease (increase) pollution to mitigate the rent of asymmetric information.

ACKNOWLEDGEMENTS

Support from the Foundation Enrico Mattei is gratefully acknowledged. I am grateful to C. Carraro, C. Crampes, J. Crémer, C. Dosi, M. Moretto and J. Tirole for their comments on the first draft.

NOTES

1. See Bohm (1970) and Laffont (1988) for a discussion of non convexities of externalities. See appendix 1 for an example.
2. This "distributional" problem could be generated differently by having a social welfare function with a lower weight on the firm's profit and no cost of social funds. The only difference with the present analysis would be that the full information pricing rule would be marginal cost pricing instead of Ramsey pricing (see Baron-Myerson (1982)).
3. Note that here $(1 + \lambda)$ is the economy wide price of public funds and is considered as fixed. It is not the multiplier of the firm's budget constraint as in the usual Ramsey-Boiteux analysis. However, a similar analysis could be performed for a firm with a budget constraint and no direct transfer from the regulator to the firm. The multiplier λ would then be a function of β.
4. If only a random signal $d_1 + d_2 + \tilde{\varepsilon}$ is observed where $\tilde{\varepsilon}$ is a random variable, the reasoning below is unchanged. Also, if actual pollution is $d_1 + d_2 + \tilde{\varepsilon}$ the analysis holds if we substitute the expectation of the marginal damage $V'(d_1 + d_2 + \tilde{\varepsilon})$.

REFERENCES

Baron, D. (1985a). "Noncooperative Regulation of a Nonlocalized Externality", *Rand Journal of Economics*, 16, pp. 553–568.

Baron, D. (1985b). "Regulation of Prices and Pollution under Incomplete Information", *Journal of Public Economics*, 28, pp. 211–231.

Baron, D., Myerson, R. (1982) "Regulating a Monopolist with Unknown Costs", *Econometrica*, 50, pp. 911–930.

Besanko, D. (1987). "Performance versus Design Standards in the Regulation of Pollution", *Journal of Public Economics*, 34, pp. 19–44.

Bohm, P. (1970), "Pollution, Purification et Théorie des Effets Externes", *Annales de l'INSEE*, 3.

Bowes, M., MAcCauley, M., Palmer, K. (1991). "Incentive Based Approaches to Regulating Toxic Substances", Resources for the Future, mimeo.

Braden, J., Bouzaher, A., Johnson, G., Miltz, D. (1989). "Optimal Spatial Management of Agricultural Pollution", *American Journal of Agricultural Economics*, 71, pp. 404–413.

Buchanan, J.M. (1969). "External Diseconomies, Corrective Taxes and Market Structure", *American Economic Review*, 59, pp. 174–176.

Carlton, D., Loury, G. (1980). "The Limitations of Pigouvian Taxes as a Long-Run Remedy for Externalities", *Quarterly Journal of Economics*, 95, pp. 559–566.

Dasgupta, P., Hammond, P., Maskin, E. (1980). "A Note on Imperfect Information and Optimal Pollution Control", 47, pp. 857–800.

Graham-Tomasi, T. (1992). "Nonpoint Pollution Regulation in a Spatial Model: Issues and Analysis", mimeo.

Graham-Tomasi, T., Segerson, K., Braden, J. (1992). "Issues in the Design of Incentive Schemes for Nonpoint Source Pollution Control", Workshop on Nonpoint Pollution Control, Foundation Mattei, Milan.

Groves, T. (1973). "Incentives in Teams", *Econometrica*, 41, pp. 617–631.

Guesnerie, R., Laffont, J.J. (1984). "A Complete Solution to a Class of Principal-Agent Problems with an Application to the Control of a Self-Managed Firm", *Journal of Public Economics*, 25, pp. 329–369.

Kwerel, E. (1977). "To Tell the Truth: Imperfect Information and Optimal Pollution Control", *Review of Economic Studies*, 44, pp. 595–602.

Laffont, J.J. (1977). *Effets Externes et Théorie Economique*, Editions du CNRS, Paris.

Laffont, J.J. (1988). *Fundamentals of Public Economics*, MIT Press, Cambridge.

Laffont, J.J., Tirole, J. (1990), "The Regulation of Multiproduct Firms", *Journal of Public Economics*, 43, pp. 1–66.

Laffont, J.J., Tirole, J. (1991a). "Provision of Quality and Power of Incentive Schemes in Regulated Industries", in *Equilibrium Theory and Applications*, J. Gabszewicz and A. Mas-Colell (Eds.), Cambridge U.P.

Laffont, J.J., Tirole, J. (1991b). "The Politics of Government Decision Making: A Theory of Regulatory Capture", *Quarterly Journal of Economics*, 106, pp. 1089–1127.

Lewis, T., Sappington, D. (1992). "Allocating Pollution Permits and Other Scarce Resource Rights", University of Florida, mimeo.

Plott, C. (1966). "Externalities and Corrective Taxes", *Economica*, 33, pp. 84–87.

Roberts, M., Spence, M. (1976). "Effluent Charges and Licenses under Uncertainty", *Journal of Public Economics*, 5, pp. 193–208.

Schulze, W., D'Arge, R.C (1974). "The Coase Proposition, Informational Constraints and Long Run Equilibrium", *American Economic Review*, 64, pp. 763–772.

Segerson, K. (1988). "Uncertainty and Incentives for Nonpoint Pollution Control", *Journal of Environmental Economics and Management*, 15, pp. 87–98.

Spulber, D.F. (1985). "Efficient Regulation and Long-Run Optimality", *Journal of Environmental Economics and Management*, 12, pp. 103–116.

Spulber, D.F. (1989). "Optimal Environmental Regulation under Asymmetric Information", *Journal of Public Economics*, 35, pp. 163–181.

Van Egteren, H. (1989). "Environmental Quality and Price Regulation in a Multi-Dimensional Monopoly Screening Model", University of Illinois, mimeo.

Xepapadeas, A. (1991). "Environmental Policy under Imperfect-Information: Incentives and Moral Hazard", *Journal of Environmental Economics and Management*, 20, pp. 113–126.

3. Controlling Environmental Externalities: Observability and Optimal Policy Rules

ANASTASIO XEPAPADEAS

University of Crete, Department of Economics, Crete, Greece

1. INTRODUCTION

The major factor which distinguishes nonpoint from point source pollution problems is our inability to directly observe individual emissions of the former or to infer them from observable inputs or from the ambient concentration of the pollutant. This has a profound impact on the choice of policy instruments for each case. For point source pollution, standard instruments like effluent taxes or emission standards can be used to induce firms to follow a socially desirable emission policy. In the nonpoint source (NPS) case, the asymmetries in information between the environmental agency and the individual polluter is a source of moral hazard. As a result, the unmonitored firm will discharge more than is socially optimal. Policy instruments developed to deal with this situation take mainly the form of tax/subsidy schemes that depend on the deviations between measured and socially desirable ambient pollutant concentration levels (Meran and Schwalbe 1987, Segerson 1988, Xepapadeas 1991, 1992).

In general, however, an NPS problem can be expected to exhibit some point source features. That is, based on a firm's specific characteristics, like for example location, inputs, or production processes used, some part of an individual firm's emissions could be observed. The information about the observed part of total emissions could, in principle, be used to levy an effluent tax. Since not all the emissions are observed, some kind of NPS or ambient tax might be required to obtain socially desirable emissions.

The purpose of this paper is to derive policy rules consisting of pairs of effluent and ambient taxes that, when applied in situations of imperfect observability of

C. Dosi & T. Tomasi (eds.), Nonpoint Source Pollution Regulation: Issues and Analysis, 67–86.

individual emissions, will lead to the socially desirable ambient concentration levels of the pollutant. The observed part of a firm's emissions might be increased if the environmental agency invested in pollution monitoring equipment. Thus, at a second level, the paper tries also to explore the conditions under which an environmental agency would be willing to undertake the cost of investing in pollution monitoring equipment in order to increase its information about individual emissions, and consequently be able to reduce or even abolish ambient taxes in favor of effluent taxes. If such an investment is worth undertaking, then the optimal policy rules will be augmented by an investment policy in monitoring equipment.

2. SOCIAL AND PRIVATE OPTIMUM

We consider a market consisting of $i = 1, ..., n$ firms that produce a homogeneous product. Output price is determined exogenously. Output production generates pollution that can be abated at a cost. The benefit of the ith firm at each instant of time can be written as a function of its discharges (Malik 1990).

$$B_i = B_i(e_i(t)), B_i \geq 0, B_i^{''} < 0 \text{ for all } t \in [0, \infty) \tag{1}$$

where $e_i \in E_i \subset R_+$ denotes discharges of the ith firm and E_i is assumed to be compact and convex. The accumulation of the pollutant is described by the differential equation

$$\dot{X}(t) = \sum_i e_i(t) - dX(t), X(0) = X_0 \geq 0 \tag{2}$$

where $X \in \boldsymbol{X} \subset R_+$ with $\boldsymbol{X}$ compact and convex, denotes the ambient concentration of the pollutant generated by the firms' productive activities and d denotes the exponential natural pollution decay rate.

Assuming a convex damage function

$$D = D(X(t)), D^{'} > 0, D^{''} > 0 \text{ for all } t \in [0, \infty) \tag{3}$$

the social optimum, from the viewpoint of a social planner solving

$$\max_{\{e_i(t)\}} \int_0^{\infty} e^{-\rho t}[\sum_i B_i(e_i) - D(X)]dt$$

subject to (3), is characterized by the following conditions (Xepapadeas 1992):

$$B_i'(e_i^*) + \lambda \leq 0, e_i^* \geq 0 \quad i = 1, ..., n \tag{4.1}$$

$$\dot{\lambda} = (\rho + d)\lambda + D'(X) \tag{4.2}$$

$$\dot{X} = \sum_i B_i(e_i^*) - dX \tag{4.3}$$

$$\lim_{t \to \infty} e^{-\rho t}\lambda(t)X(t) = 0 \tag{4.4}$$

where $\lambda(t) < 0$ is the dynamic social shadow cost of pollutant accumulation and ρ is the social discount rate.

From (4.1) we obtain for interior solutions $e_i^* = e_i(\lambda)$ as the socially desired emission level for firm i. Substituting e_i^* in (4.3) and solving the system of (4.2), (4.3) we obtain the socially desired ambient concentration path, $X^*(t)$, for the pollutant and the path for the pollutant's social shadow cost $\lambda^*(t)$. By standard arguments under perfect observability of individual emissions, an effluent tax set as $\tau(t) = -\lambda^*(t)$ could secure socially optimal discharge and accumulation levels.

Let $a_i(t)$ represent the part of emissions by firm i that is observed at time t, that is $a_i \in [0, e_i]$. It is assumed that firm i's observed emissions depend on the following factors: (i) the existing stock of capital in pollution monitoring equipment at time t, $K(t)$; (ii) a parameter z_i that reflects specific characteristics of the firm; and (iii) the actual emissions of the firm at time t, $e_i(t)$. Thus the observed emissions are assumed to be determined according to the function:

$$a_i(t) = a_i(K(t), z_i, e_i(t)) \tag{5}$$

The following assumptions are made about function a_i:

$$(a_z, a_K, a_e) > 0, (a_{zz}, a_{KK}) < 0, (a_{ee}, a_{eK}, a_{zK}) \geq 0$$

where $K \in \boldsymbol{K}$, $z_i \in \boldsymbol{Z}_i$, $\boldsymbol{K}, \boldsymbol{Z}_i \in R_+$ assumed convex and compact. Therefore higher values of z_i are assumed to reflect superior conditions of observability of firms' emissions, and the more a firm discharges, the higher the observable part of its emissions are.

If $K = 0$, then the values $a_i(0, z_i, e_i) \geq 0$ correspond to that part of emissions that can be observed without any monitoring equipment, but solely on the basis of firm i's specific characteristics. It is assumed that some value $\bar{z}_i$ exists such that $a_i(0, z_i, e_i) = 0$ for $z_i < \bar{z}_i$; and also that a value K^0 exists such that for all i, $a_i(K, z_i, e_i) \to e_i$ as $K \to K^0$ and that $a_{iK}(K^0, z_i, e_i) = 0, i = 1, ..., n$.

Assuming no uncertainty and no measurement errors, the agency observes a vector of individual emissions, $\underline{a} = (a_1, ..., a_n)$ and imposes an effluent tax τ_i on those emissions. The agency also continuously observes the ambient concentration of the pollutant $X(t)$ at some receptor point. Given the information on total individual emissions $\Sigma_i a_i$, the agency uses the accumulation process (2) to calculate the concentration $X_a(t)$ that corresponds to emissions $\Sigma_i a_i$. If $X(t) > X_a(t)$, then actual emissions (Σe_i) exceed observed emissions (Σa_i). Since under imperfect monitoring firms have an incentive to discharge more than socially desirable, therefore, $\Sigma_i e_i \geq \Sigma_i e_i^*$. If in fact $X(t) > X^*(t)$, the agency supplements its effluent tax scheme by an ambient tax b_i on observed deviations between the measured levels $X(t)$ and the desired level $X^*(t)$. The tax is imposed on every firm for which it cannot be proved by other means (e.g. continuous inspection) that $a_i = e_i^*$. The tax policy rule is thus a pair with the following properties:

$$(\tau_i, b_i) = \left\{ \begin{array}{c} \tau_i = 0, b_i > 0 \text{ if } a_i = 0, X > X^* \\ \tau_i, b_i > 0 \text{ if } a_i > 0, X > X^*, X > X_a \\ \tau_i > 0, b_i = 0 \text{ if } X = X^* \text{ or if } a_i = e_i^* \end{array} \right\}$$

In this context, the cash flow for the ith firm at each time t takes the form:

$$B_i(e_i(t)) - \tau_i a_i(t) - b_i(X(t) - X^*(t)) \tag{6}$$

The objective of the firm is to choose the control sequence $\{e_i(t)\}$ to maximize the present value of (6) over an infinite time horizon, which is subject to the transition equation (2). Since the ith firm's objective function depends on the rest of the firms' emission policy through (2), the solution to the firm's problem is carried out in the context of an n-player, non-cooperative dynamic game. It is assumed that firm i's information structure is feedback (Basar and Olsder 1982), thus $e_i = \theta_i(X(t), t)$. As is well-known, a feedback Nash equilibrium (FBNE), under which each firm can observe the concentration of the pollutant and condition its discharges at this level, is subgame perfect (Fershtman 1987). The conjecture function for each firm is further specified (Xepapadeas 1992) as:

$$\text{(CF)} \quad e_j = \bar{e}_j + \theta(X), \theta' < 0 \quad j = 1, ..., n$$

The negative slope reflects the assumption that each firm expects other firms to reduce their emissions if accumulation of the pollutant increases. Under these assumptions, the firm's problem is:

$$\max_{\{e_i(t)\}} \int_0^{+\infty} e^{-rt}[B_i(e_i) - \tau_i a_i - b_i(X - X^*)]dt$$

$$s.t.\ \ \dot{X} = e_i + \sum_{j \neq i}(\bar{e}_j + \theta(X)) - dX$$

$$a_i \leq e_i \qquad \text{(P.1)}$$

The current value Lagrangean for problem (P.1) can be written for $i = 1, ..., n$ as:

$$\mathcal{L}_i(X, e_i, ..., e_n, \mu_i) = B_i(e_i) - \tau_i a_i - b_i(X - X^*) + \mu_i[e_i + \textstyle\sum_{j \neq i}(\bar{e}_j + \theta(X)) - dX] + \psi_i[e_i - a_i(K, z_i, e_i)]$$

where $\mu_i < 0$ is the dynamic private shadow cost of pollutant concentration, and r is the common private discount rate.

The necessary conditions for FBNE are:

$$\frac{\partial H}{\partial e_i} = B_i^{'}(e_i) - \tau_i \frac{\partial a_i}{\partial e_i} + \psi_i(1 - \frac{\partial a_i}{\partial e_i}) + \mu_i \leq 0, e_i \geq 0\ i = 1, ..., n \qquad (7.1)$$

$$\dot{\mu}_i = [r + d - (n-1)\theta^{'}(X)]\mu_i + b_i \quad i = 1, ..., n \qquad (7.2)$$

$$\dot{X} = \sum_{i=1}^{n} e_i - dX \qquad (7.3)$$

$$\mathcal{L}_{\psi_i} \geq 0, \psi_i[e_i - a_i(K, z_i, e_i)] = 0, \psi_i \geq 0 \qquad (7.4)$$

along with the Arrow-type transversality conditions $\exp(-rt)\mu_i(t)X(t) = 0$ as $t \to \infty$. Conditions (7.1)-(7.4) characterize the private optimum and can be used to analyze policy rules consisting of the pairs (τ_i, b_i) and to compare the outcomes of private markets with the social optimum.

If observability is perfect, that is $a_i = e_i$, then $\partial a_i / \partial e_i = 1$ and $b_i = 0$. Thus the ambient concentration of the pollutant does not enter the firm 's problem. This means of course that problem (P.1) is not constrained by the transition equation (2). Therefore $\mu_i = 0$ and (7.1) becomes $B'(e_i) = \tau_i$, which for $\tau_i = -\lambda$ is the effluent fee solution under perfect monitoring.

If observability is not perfect, $a_i < e_i$ and by (7.4) $\psi_i = 0$. Then (7.1) becomes for interior solutions:

$$B^{'}(e_i) - \tau_i \frac{\partial a_i}{\partial e_i} + \mu_i = 0 \qquad (8)$$

Solving for e_i, the short-run emission function consistent with the private optimum is:

$$\hat{e}_i = e_i(K, \tau_i, z_i, \mu_i) \qquad (9)$$

Using the implicit function theorem, the following short-run comparative statics are obtained from (8), where subscript i has been dropped for simplicity and remaining subscripts denote partial derivatives:

$$\frac{\partial e}{\partial K} = \frac{\tau a_{eK}}{B_{ee} - \tau a_{ee}} < 0, \frac{\partial e}{\partial \tau} = \frac{a_e}{B_{ee} - \tau a_{ee}}, \frac{\partial e}{\partial z} = \frac{\tau a_{ez}}{B_{ee} - \tau a_{ee}} < 0,$$
$$\frac{\partial e}{\partial \mu_i} = -\frac{1}{B_{ee} - \tau a_{ee}} > 0 \tag{9.1}$$

That is, an increase in the stock of capital in monitoring equipment, the effluent tax rate or the information about firms' specific characteristics will decrease individual profit maximizing emissions. Since $\mu_i < 0$, a reduction in the absolute value of the private shadow cost of the pollutant will increase private emissions. By substituting $\hat{e}_i$ from (9) in (7.3), we have a system of $(n + 1)$ differential equations for the private shadow costs of the pollutant μ_i and the pollutant's accumulation X. This system can be used to analyze some properties of the long-run (steady state) equilibrium for μ_i and X.

Proposition 1 *If a unique steady state for the system (7.2), (7.3) with $e_i \equiv \hat{e}_i$ defined as:*

$$(\mu_1^\infty, ..., \mu_n^\infty, X^\infty) : \dot{\mu}_1 = ... = \dot{\mu}_n = \dot{X} = 0$$

exists, then the following steady state comparative statics holds:

$$\left(\frac{\partial X^\infty}{\partial K}, \frac{\partial X^\infty}{\partial \tau_i}, \frac{\partial X^\infty}{\partial z_i}, \frac{\partial X^\infty}{\partial b_i}\right) < 0$$

$$\left(\frac{\partial \mu_i^\infty}{\partial K}, \frac{\partial \mu_i^\infty}{\partial \tau_i}, \frac{\partial \mu_i^\infty}{\partial z_i}, \frac{\partial \mu_i^\infty}{\partial b_i}\right) \leq 0 \quad i = 1, ..., n$$

Proof: see Appendix.

Therefore, an increase in the stock of monitoring equipment, the tax parameters or the information parameters will reduce the steady state pollutant accumulation, and will not reduce the private cost of pollutant accumulation at the steady state.

The results obtained in this section characterize the social optimum and the private optimum when firms face a dual tax scheme consisting of an effluent and an ambient tax. The optimality conditions derived will be used in the following section to derive the optimal tax scheme.

3. REGULATION WITHOUT INVESTMENT IN POLLUTION MONITORING EQUIPMENT

In this section, the conditions that determine the optimal pair of effluent-ambient tax are derived. It is assumed that the agency does not invest in pollution monitoring equipment, but tries to control the externality by using the costless information z_i, and a given stock of capital in monitoring equipment $K \geq 0$ that is assumed to have been inherited without cost from the past.

Combining (8) with (4.1), the tax policy parameters that will reproduce the social optimum should satisfy the relationship:

$$\tau_i \frac{\partial a_i}{\partial e_i}(K, z_i, e_i) - \mu_i = -\hat{\lambda}^* \quad i = 1, ..., n \tag{10}$$

where $\hat{\lambda}^* = \lambda(t; \rho, d)$ is the pollutant's social shadow cost obtained as described in the previous section from the solution of (4.2), (4.3).

System (10) can be used to determine the optimal tax policy pairs (τ_i, b_i). Provided that a unique solution to the system of $(n + 1)$ differential equations (7.2) and (7.3), after replacing e_i with $\hat{e}_i$ from (9), exists, the solution will depend on the parameters of the system (e.g., Brock and Malliaris 1989). The solution will have the form:

$$\hat{\mu}_i = \mu_i(K, \underline{\tau}, \underline{b}, \underline{z}; t), \quad i = 1, ..., n \tag{11.1}$$

$$\hat{X} = X(K, \underline{\tau}, \underline{b}, \underline{z}; t) \tag{11.2}$$

where $\underline{\tau} = (\tau_1, ..., \tau_n), \underline{b} = (b_1, ..., b_n), \underline{z} = (z_1, ..., z_n)$.

Substituting (11.1) into (10), we have the system:

$$\tau_i \frac{\partial a_i}{\partial e_i}(K, z_i, e_i(K, \tau_i, z_i, \mu_i(K, \underline{\tau}, \underline{b}, \underline{z}; t))) - \mu_i(K, \underline{\tau}, \underline{b}, \underline{z}; t) + \hat{\lambda}^* = 0$$
$$i = 1, ..., n \tag{12}$$

If the Jacobian of this system with respect to b does not vanish, the implicit function theorem can be used to obtain solutions of the form:

$$\hat{b}_i = b_i(K, \underline{\tau}, \underline{z}, \hat{\lambda}^*; t), \quad i = 1, ..., n \tag{13}$$

Relation (13) indicates that the optimal tax policy pair $(\hat{b}_i, \hat{\tau}_i)$ can be obtained by choosing arbitrary values for the effluent tax rates $\hat{\tau}_i$ and then using (12) to obtain the optimal tax on the deviations from the desired path of pollution accumulation.

The indeterminacy of policy function (13) can be resolved by setting the effective tax rates τ_i at the social welfare maximizing level λ^*. That is, the effective tax rates are set equal to marginal damages as reflected by the pollutant's shadow cost. The ambient taxes are then obtained by (13) as:

$$\hat{b}_i = b_i(K, \lambda^*, \underline{z}; t)$$

By (14) the optimal ambient tax depends on the state of monitoring equipment, the social shadow cost of the pollutant and the vector of information parameters. The effects on $\hat{b}_i$ from changes in K or z_i are presented in the following proposition.

Proposition 2 *If $\partial^2 a_i/\partial e_i^2 = 0$ or sufficiently small, then $(\partial\hat{b}_i/\partial K, \partial\hat{b}_i/\partial z_i) <$ 0.*

Proof: see Appendix.

This proposition implies that for firms with specific characteristics that allow better observation of their emissions, the ambient tax will be lower, and also that an increase in the stock of monitoring equipment allows for a reduction in the ambient tax for all firms.

Given, therefore, some positive, observed part of firms' emissions, the agency can obtain the desired behavior by charging an effluent tax for the observed emission, equal to the social shadow cost of the pollutant, and an ambient tax that depends on the effluent fee itself, the stock of monitoring equipment and the specific characteristics of firms. If, on the other hand, $a_i = 0, (z_i < \hat{z}_i)$ the problem is pure NPS and $\tau_i = 0$. Then $B_i' = -\mu_i$ and optimality requires $\mu_i = \lambda^*$. Proceeding as before, $\hat{b}_i$ can be obtained as the solution of the system $\mu_i(\underline{b}, K, \underline{z}, : t) = \lambda^*$. In Xepapadeas (1992), the ambient tax is chosen so as to achieve the social optimum at steady state. If, however, b_i is chosen as above, the pollutant accumulation path resulting from profit maximization coincides with the socially optimal path.

4. REGULATION WITH INVESTMENT IN POLLUTION MONITORING EQUIPMENT

The results obtained in the previous section suggest that under incomplete observability of individual emissions, dual instruments in the form of ambient taxes and effluent taxes can be used, and that the optimal ambient tax rates are reduced if a larger stock of capital in monitoring equipment allows better observability of individual emissions.

The important question, however, is whether increased observability is desirable, since the combination of instruments developed so far can achieve the social optimum and since the acquisition of monitoring equipment will incur additional costs to the environmental agency. That is, if we can achieve the desired emission levels by using a combination of effluent and ambient taxes, based on whatever information about individual emissions that can be obtained without cost, is there any reason for investing in pollution monitoring equipment in order to increase our information about individual emissions and consequently reduce ambient taxes? The possibility of giving an affirmative answer to the above question depends on how important certain factors like the following might be.

(i) It is very likely that the introduction and implementation of ambient taxes would be difficult. This is because while individual dischargers may be willing to pay an effluent fee, it is very doubtful whether they will be ready to accept an ambient tax, since their payments under this tax will strongly depend on the actions of other firms. The higher the rates are, the stronger the opposition to this type of tax might be, which in addition would tend to be discriminatory. This calls for investment in monitoring equipment in order to increase observability and reduce the ambient taxes.

(ii) Under the type of ambient tax schemes developed here, the total payments on a marginal unit of pollution exceed the marginal damage caused by the unit. This is a result of the non-balanced budgeting nature of the incentive scheme (Segerson, 1988, Xepapadeas, 1991). This of course implies that ambient taxes impose on firms costs that exceed the damage cost of marginal emissions. Therefore, reduction in ambient taxes will reduce this type of extra costs.

The above factors suggest that there could be extra benefits associated with investment in pollution monitoring equipment, which allows the regulator to increase the observed part of firms emissions and to reduce ambient tax rates. Without specifying exactly the way that an increase in the stock of pollution monitoring equipment generates benefits, let us assume that such benefits can be captured by a function $U(a_1(t), ...a_n(t))$.

It is assumed that $\partial U/\partial a_i \geq 0$, that U has a negative definite Hessian and that $\partial U(a_i, \underline{a}_{-i})/\partial a_i \rightarrow 0$ if $a_i \rightarrow e_i$ and $\underline{a}_{-i} = e_{-i}$ where $\underline{a}_{-i} = (a_1.., a_{i-1}, a_{i+1}, ..a_n)$. In this context, the agency tries to choose optimal taxes $\tau_i(t), b_i(t)$, an investment policy $I(t)$ and emission levels $e_i(t)$ to maximize social welfare. Since, however, investment incurs costs and observability of emissions generates benefits, social welfare at each instant of time is defined as:

$$\sum_i B_i(e_i) - D(X) - C(I) + U(\underline{a}) \tag{14}$$

where $C(I)$ includes full investment adjustment costs $C(0) = 0, C'(I) > 0, C''(I) > 0$.

In choosing its instruments, the agency is constrained by the need to choose emission levels in such a way that if firms follow its instructions on their emission policy, they will maximize the present value of their profits.[1] In this context, the problem for the agency can be written as:

$$\max_{\{e_i(t), \tau_i(t), b_i(t), I(t)\}} \int_0^{+\infty} e^{-\rho t}[\sum_i B_i(e_i) - D(X) - C(I) + U(\underline{a})]dt \tag{15.1}$$

$$s.t. \quad \dot{K} = I - \delta K, K(0) = K_0 \geq 0 \tag{15.2}$$

$$e_i(t) \in \arg \max_{\{e_i(t)\}} \int_0^{+\infty} e^{-rt}[B_i(e_i) - \tau_i a_i(K, z_i, e_i) - b_i(X - X^*)]dt, \forall\, i \tag{16.1}$$

$$s.t. \quad \dot{X} = e_i + \sum_{j+i} \bar{e}_j + \theta(X)) - dX, X(0) = X_0 \geq 0 \tag{16.2}$$

$$a_i \leq e_i \tag{16.3}$$

In the above problem, constraint (15.2) reflects capital accumulation, with $\delta > 0$ being the exponential depreciation rate. Constraints (16.1), (16.2) and (16.3) reflect the individual firm's optimizing behavior. Substituting (16.1), (16.2) and (16.3) with the corresponding first order conditions, the current value Lagrangean for the agency becomes:

$$\mathcal{L} \ = \sum_i B_i(e_i) - D(X) - C(I) + U(\underline{a}) + v(I - \delta K) +$$

$$\sum_i \gamma_i(B_i' - \tau_i \frac{\partial a_i}{\partial e_i} + \psi_i(1 - \frac{\partial a_i}{\partial e_i}) + \mu_i) +$$
$$\sum_i u_i[((r + d) - (n - 1)\theta'(X))\mu_i + b_i] + \zeta(\sum_i e_i - dX) \quad (17)$$

In this expression, $v > 0$ reflects the social value of investment in pollution monitoring equipment, u_i the value that society puts on the private cost of the pollutant's accumulation, while $\zeta < 0$ is the social cost of the pollutant's accumulation which is defined in terms of the objective function (14).

Assuming interior solutions and $\psi_i = 0$, that is, observability is not perfect, the first order conditions for the choice of the controls imply the following for $i = 1, ..., n$:

$$B_i' + \zeta + \frac{\partial U}{\partial a_i}\frac{\partial a_i}{\partial e_i} = 0 \quad (18.1)$$

$$C'(I) = v \quad (18.2)$$

$$\gamma_i = u_i = 0 \quad (18.3)$$

$$\mathcal{L}_{\gamma_i} \geq 0, B_i' - \tau_i \frac{\partial a_i}{\partial e_i} + \mu_i = 0 \quad (18.4)$$

The canonical equations are defined as:

$$\dot{v} = (\rho + d)v - \sum_i \frac{\partial U}{\partial a_i}\frac{\partial a_i}{\partial K} \quad (19.1)$$

$$\dot{K} = I - \delta K \quad (19.2)$$

$$\dot{\zeta} = (\rho + d)\zeta + D'(X) \quad (20.1)$$

$$\dot{X} = \sum_i e_i - dX \quad (20.2)$$

$$\dot{\mu}_i = (r + d - (n - 1)\theta'(X))\mu_i + b_i \quad (20.3)$$

The Arrow-type transversality conditions are also assumed to be satisfied at the solution.

The above conditions can be used to analyze optimal policy rules. Relation (19.1) in conjunction with (4.2) implies that $\zeta \equiv \lambda$, therefore the condition for optimal emissions becomes:

$$B'(e) + \lambda + \frac{\partial U}{\partial a_i}\frac{\partial a_i}{\partial e_i} = 0 \quad (21)$$

or combined with (18.4):

$$-\lambda - \frac{\partial U}{\partial a_i}\frac{\partial a_i}{\partial e_i} = \tau_i \frac{\partial a_i}{\partial e_i} - \mu_i \tag{22}$$

which implies that $\hat{e}_i = e_i(K, \mu_i, \lambda, z_i, \tau_i)$ with

$$\frac{\partial e_i}{\partial K} = -\frac{U_{aa}a_K a_e + a_{eK}[U_a + \tau_i]}{U_{aa}a_e^2 + a_{ee}[U_a + \tau_i]} \leq 0 \tag{23}$$

provided that a_{ee}, a_{eK} are either zero or sufficiently small.[2] The rest of the optimality conditions become:

$$\dot{v} = (\rho + d)v - \sum_i \frac{\partial U}{\partial a_i}\frac{\partial a_i}{\partial K} \tag{24.1}$$

$$\dot{K} = h(v) - \delta K \tag{24.2}$$

$$\text{where } h(v) \text{ is the solution of } C'(I) = v \text{ with } h'(v) > 0 \tag{24.3}$$

$$\dot{X} = \sum_i e_i(K, \mu_i, \tau_i, z_i) - dX \tag{25.1}$$

$$\dot{\lambda} = (\rho + d)\lambda + D'(X) \tag{25.2}$$

$$\dot{\mu}_i = (r + d - (n-1)\theta'(X))\mu_i + b_i \tag{26}$$

Solving (24.1) for v and (26) for μ_i at the steady state and substituting into (24.2) and (25.1) respectively, we obtain:

$$\dot{K} = h(v(K)) - \delta K \tag{27.1}$$

$$\dot{X} = \sum_i e_i(K, \mu_i(X), \tau_i, z_i) - dX \quad \mu_i'(X) < 0 \tag{27.2}$$

The properties of the steady state for this system, defined as $(X^*, K^*) : \dot{X} = \dot{K} = 0$, are summarized in Proposition 3.

Proposition 3 *If* $(\partial a_i/\partial K) > (\partial a_i/\partial e_i)(\partial e_i/\partial K)$, *then the steady state* X^*, K^* *is a stable node.*[3]

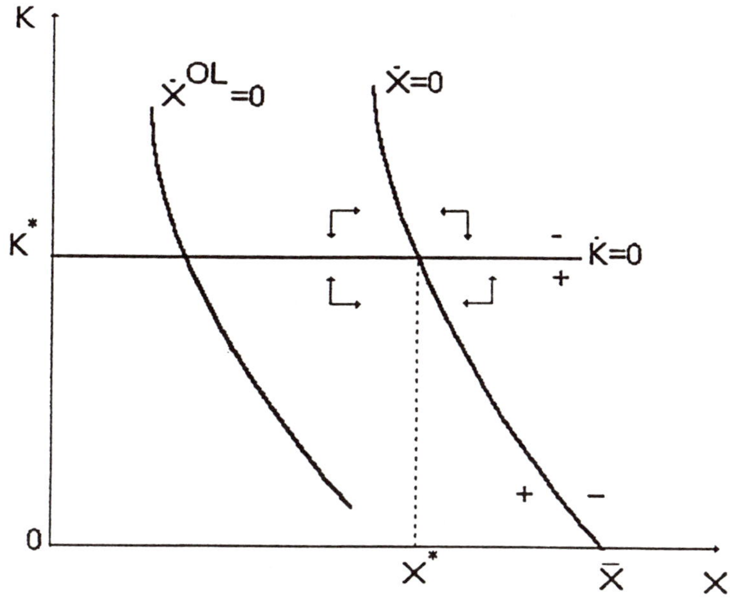

Fig. 1. Steady state stocks of pollution and capital

Proof: see Appendix.

The steady state is presented in Figure 1. As shown in this figure, the effect of positive capital in monitoring equipment reduces the pollutant's steady state accumulation level. In the same figure, $\bar{X}$ corresponds to the case where no investment in monitoring equipment is carried out.

The optimality conditions derived above can also be used to explore the effects of the dischargers' information structure. If $\theta'(X) = 0$, then the firms' information structure is said to correspond to an open loop Nash equilibrium (OLNE). The information structure is said to be open loop if $e_i = \theta_i(X(0), t)$. The open loop Nash equilibrium is not subgame perfect. From (20.3) at the steady state $\mu_i^{OL} < \mu_i^{FB}$ and this implies that $e_i^{OL} < e_i^{FB}$. So in Figure 2, the $\dot{X} = 0$ locus shifts inside under the open loop information structure to the $\dot{X}^{OL} = 0$ locus. Thus in OLNE, the same steady state pollutant accumulation level as the FBNE can be obtained with less investment in monitoring equipment.

This is a result of the negative slope assumption of the conjecture function. If a firm expects other firms to reduce their emissions when the ambient pollution level increases, it has an incentive to discharge more, since it expects the increased emissions to be partly offset by the reduced emissions of others. When everybody behaves in this way, as is the case under a feedback information structure, emissions go up as compared to the open loop information structure, where firms do not condition their emissions on the basis of ambient concentration levels. Therefore, in FBNE more capital in monitoring equipment is required to obtain the same ambient levels as in OLNE.

The solution $\hat{X}(t)$, $\hat{K}(t)$ of system (27) can be substituted back into (24.1), (25.2) and (26) to obtain $\hat{v}(\underline{\tau}, \underline{b}, \underline{z}; t)$, $\hat{\lambda}(\underline{\tau}, \underline{b}, \underline{z}; t)$, $\hat{\mu}(\underline{\tau}, \underline{b}, \underline{z}; t)$. These solutions will, in turn, determine the optimal policy rule $(\hat{\tau}_i, \hat{b}_i, \hat{I}_i)$ through the following system obtained by (22) and (18.2):

$$-\hat{\lambda}(\underline{\tau}, \underline{b}, \underline{z}; t) - \frac{\partial U}{\partial a_i}\frac{\partial a_i}{\partial e_i}(\underline{\tau}, \underline{b}, \underline{z}; t) = \tau_i \frac{\partial a_i}{\partial e_i}(\underline{\tau}, \underline{b}, \underline{z}; t) - \hat{\mu}_i(\underline{\tau}, \underline{b}, \underline{z}; t) \forall\ i \quad (28.1)$$

$$C'(I) = \hat{v}((\underline{\tau}, \underline{b}, \underline{z}; t) \quad (28.2)$$

The indeterminacy in the solution for $(\underline{\tau}, \underline{b})$ in system (28.1) can be resolved by substituting τ_i with λ^*, the pollutant's shadow cost, corresponding to the first best social optimum, and then solving for b_i. The optimal investment policy is then determined through (28.2).

5. CONCLUDING REMARKS

This paper explores the possibilities of developing policy rules for controlling environmental externalities when the observability of individual firms' emission processes can be improved either by using information on the specific characteristics of firms or by investment in pollution monitoring equipment. In this context, the policy rules consist in general of an effluent tax on the observable part of the firms' emissions, of an ambient or NPS tax on deviations from the optimal path of pollutant accumulation, and an investment policy on monitoring equipment. Ambient tax rates are discriminatory, in the sense that the higher the observable proportion of firms' emissions, the lower the NPS tax rate that the firm pays.

If the environmental agency chooses not to invest at all but relies on individual characteristics to infer emissions by individual firms, then the optimal policy consists of a combination of an effluent tax and an ambient tax. The effluent tax will be zero for the specific firms whose emissions cannot be observed at all. These firms will pay a relatively high ambient tax, since the lower the observed part of its emissions, the higher the ambient tax. This policy rule would be optimal in the sense that the path of pollutant accumulation, resulting when profit maximizing firms are subjected to the policy, is the same as the socially desired path.

The agency will invest in pollution monitoring equipment if increased observability of individual firms' emissions, and therefore lower ambient taxes, generates benefits. These benefits can be associated either with political economy issues of ambient taxes or with the excess cost (over marginal damages) of non-balanced budgeting ambient taxes.

In this context, the policy rule consists of a pair of effluent and ambient taxes and an investment policy in monitoring equipment. The steady state capital stock of pollution monitoring equipment will be positive, and the steady state pollutant accumulation level will be lower than it would be without pollution monitoring equipment. It should be noted finally, that the agency will engage in investment in pollution monitoring only if observability of individual emissions generates benefits. If, however, $U(\underline{a}) \equiv 0$, then $v = 0$ and $K = 0$. The agency will observe any emissions possible, according to individual characteristics, without incurring any costs and will treat all unobservable items as an NPS problem.

APPENDIX. PROOFS

(i) Proposition 1

From the system of (7.2) at the steady state, we obtain

$$0 = [r + d - (n-1)\theta'(X)]\mu_i + b_i \quad i = 1, ..., n \tag{A.1}$$

$$\text{with } \frac{\partial \mu_i}{\partial X}|_{\dot{\mu}_i=0} = -\frac{b_i(n-1)\theta''(X)}{[r+d-(n-1)\theta''(X)]^2} \leq 0 \text{ for } \theta''(X) \geq 0 \tag{A.2}$$

Substituting μ_i from (A.1) into (7.3) at the steady state we obtain

$$0 = \sum_1 e_i\left(K, \tau_i, z_i, \frac{-b_i}{r+d-(n-1)\theta'(X)}\right) - dX \tag{A.3}$$

$$X^\infty = X(K, \underline{\tau}, \underline{z}, \underline{b}, r, d, n) \tag{A.4}$$

Substituting (A.4) into (A.1) we obtain

$$\mu_i^\infty = \mu_i(K, \underline{\tau}, \underline{z}, \underline{b}, r, d, n) \tag{A.5}$$

Relations (A.4) and (A.5) indicate the steady state levels of pollutant accumulation and pollutant's private shadow cost as functions of the stock of monitoring equipment K, the tax parameters $(\underline{\tau}, \underline{b})$ and the information parameters $\underline{z}$ along with the rest of the parameters of the problem.

Using the short run comparative static results (9.1) and relations (A.2), we can obtain from (A.3) and (A.1) the steady state comparative statics as:

$$\frac{\partial X^\infty}{\partial K} = \frac{1}{D_1}\left(-\frac{\partial e_i}{\partial K}\right) < 0, \frac{\partial X^\infty}{\partial \tau_i} = \frac{1}{D_1}\left(-\frac{\partial e_i}{\partial \tau_i}\right) < 0,$$

$$\frac{\partial X^\infty}{\partial z_i} = \frac{1}{D_1}\left(-\frac{\partial e_i}{\partial z_i}\right) < 0, \frac{\partial X^\infty}{\partial b_i} = \frac{1}{D_1}\left(-\frac{\partial e_i}{\partial \mu_i}\frac{\partial \mu_i}{\partial b_i}\right) < 0$$

$$\text{where } D_1 = \sum_i \frac{\partial e_i}{\partial \mu_i}\frac{\partial \mu_i}{\partial X} - d < 0 \tag{A.6}$$

$$\frac{\partial \mu^{\infty_i}}{\partial K} = \frac{1}{D_2}\left((n-1)\theta''\frac{\partial X^\infty}{\partial K}\right) \leq 0, \frac{\partial \mu_i^\infty}{\partial \tau_i} = \frac{1}{D_2}\left((n-1)\theta''\frac{\partial X^\infty}{\partial \tau_i}\right) < 0,$$

$$\frac{\partial \mu_i^\infty}{\partial z_i} = \frac{1}{D_2}\left((n-1)\theta''\frac{\partial X^\infty}{\partial z_i}\right) \leq 0, \frac{\partial \mu_i^\infty}{\partial b_i} = \frac{1}{D_2}\left((n-1)\theta''\frac{\partial X^\infty}{\partial b_i}+1\right) \leq 0,$$

$$\text{where } D_2 = r + d - (n-1)\theta''(X^\infty) > 0 \tag{A.7}$$

(ii) Proposition 2

By the implicit function theorem, we obtain from (12)

$$\frac{\partial b_i}{\partial h_i} = \frac{\tau_i\left[\frac{\partial^2 a_i}{\partial e_i \partial h_i} + \frac{\partial^2 a}{\partial e_i^2}\left(\frac{\partial e_i}{\partial h_i} + \frac{\partial e_i}{\partial \mu_i}\frac{\partial \mu_i}{\partial h_i}\right)\right] - \frac{\partial \mu_i}{\partial h_i}}{\frac{\partial \mu_i}{\partial b_i}\left(1 - \tau_i\frac{\partial^2 a_i}{\partial e_i^2}\frac{\partial e_i}{\partial \mu_i}\right)} \tag{A.8}$$

$$h_i = K \text{ or } z_i$$

In this expression, all the derivatives are defined, with the exception of those of μ_i with respect to K, z_i, or b_i. These are comparative dynamic derivatives, since they indicate the change in the entire optimal path of μ_i from a change of the stock of monitoring equipment or the information parameters. The signs of these derivatives can be obtained following Oniki's (1973) method. The solutions (11.1) and (11.2) can be substituted back into the system of (7.2) and (7.3) to obtain the identities:

$$\dot{\mu}_i \equiv [(r+d) - (n-1)\theta'(X(K,\underline{\tau},\underline{b},\underline{z};t))]\mu_i(K,\tau,\underline{b},\underline{z};t) + b_i \tag{A.9.1}$$

$$\dot{X} \equiv \sum_i e_i(\tau_i, z_i, \mu_i(K,\underline{\tau},\underline{b},\underline{z};t)) - dX(K,\underline{\tau},\underline{b},\underline{z};t) \tag{A.9.2}$$

Considering perturbations in z_i, the following system of variational differential equations is obtained by differentiating system (A.9):

$$(\dot{\mu}_i)_{z_i} = -(n-1)\theta'' X_{z_i}\mu_i + [(r+d) - (n-1)\theta'(X)](\mu_i)_{z_i}\ i = 1,..,n \tag{A.10.1}$$

$$(\dot{X})_{z_i} = \sum_i\left(\frac{\partial e_i}{\partial z_i} + \frac{\partial e_i}{\partial \mu_i}(\mu_i)_{z_i}\right) - d(X)_{z_i}, \frac{\partial e_i}{\partial z_i} = 0, \tag{A.10.2}$$

$$\text{where } (\mu_i)_{z_i} \equiv \frac{\partial \mu_i}{\partial z_i}, (\dot{\mu}_i)_{z_i} \equiv \frac{\partial \dot{\mu}_i}{\partial z_i}, (X)_{z_i} \equiv \frac{\partial X}{\partial z_i}, (\dot{X})_{z_i} \equiv \frac{\partial \dot{X}}{\partial z_i}$$

Considering all μ_j, except one, as constants, the optimal perturbed arcs for $(X)_{z_i}, (\mu_i)_{z_i}$ can be studied on the plane. The perturbed terminal conditions are determined by the steady state comparative static derivatives of Proposition 1 as $(X^\infty)_{zi} < 0, (\mu_i^\infty)_{zi} < 0$. Since X(0)=X$_0$ is fixed, the perturbed initial

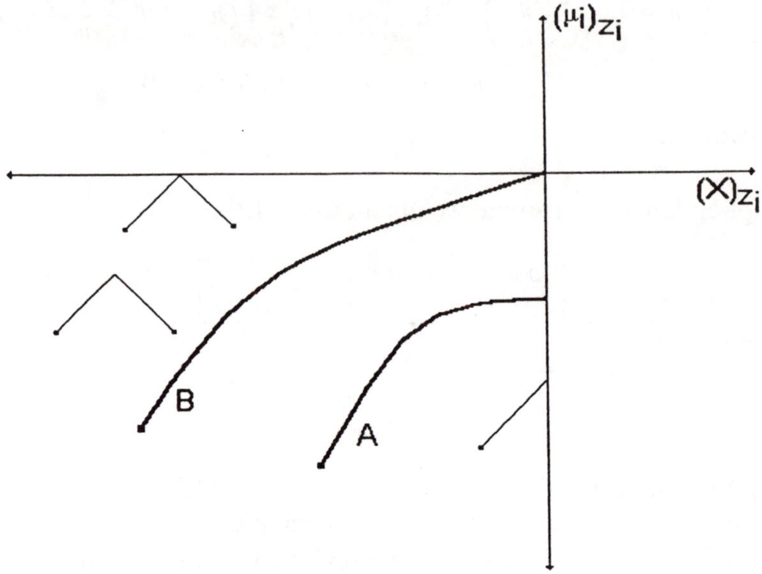

Fig. 2. Comparative dynamics

condition is $(X_0)_{z_i} = 0$. Furthermore, for the perturbed initial condition for μ_i, it is reasonable to assume that an increase in the information parameter z_i will not reduce the private shadow cost of the pollutant. Thus $(\mu_i(0))_{zi} \leq 0$.

Possible optimal perturbed arcs are presented in Figure 2. By the initial conditions, the arcs should start on the non-positive vertical axis, while by the terminal conditions, they should end up in the south-west quadrant. On arcs like A, B

$$(X)_{z_i} < 0, (\mu_i)_{z_i} < 0 \text{ for all } t$$

Similar types of results can be obtained by considering perturbations in K or b_i. Using these results in (A.8) it can be seen that $\partial b_i/\partial h_i < 0, h_i = K$ or z_i for sufficiently small $\partial^2 a_i/\partial e_i^2$.

(iii) Proposition 3

The Jacobian of system (27) around the steady state equilibrium point is defined as:

$$J_1 = \begin{pmatrix} h'v'(K) - \delta & 0 \\ \sum_i \frac{\partial e_i}{\partial K} & \sum_i \frac{\partial e_i}{\partial \mu_i}\mu'(X) - d \end{pmatrix}$$

where $v'(K) = U_{aa}(a_K + a_e e_K)a_K + U_a(a_{KK} + a_{Ke}e_K) < 0$ if $a_K > a_e e_K$. Then we have that $\det(J_1) > 0$ and $\operatorname{tr}(J_1) < 0$, thus there are two negative characteristic roots and the steady state is a stable node.

ACKNOWLEDGEMENTS

An earlier version of this paper was presented at the Workshop on Nonpoint Source Pollution Regulation sponsored by the Fondazione ENI Enrico Mattei, Milan, April 1992. I am grateful to Jean-Jacques Laffont, Kathleen Segerson, Theodore Graham-Tomasi and all the participants of this workshop for their valuable comments.

NOTES

1. No revenue constraints for the agency or maximum liability constraints for the firms are assumed in order to simplify the exposition.
2. This condition rules out the possibility that $\partial e_i/\partial K > 0$, which can be the case if a_i is strongly convex in e, for given K and z_i. If $a_{ee} > 0$ and sufficiently large, then since observability is desirable and increases with emissions, optimal emissions might increase in order to increase observability. It should be noted that the "sufficiently small" or "zero" condition on a_{ee} was also sufficient to ensure the validity of Proposition 2.
3. The stability conditions require that the absolute value of the direct increase in the observed part of emissions due to an increase in monitoring equipment $(\partial a_i/\partial K)$, exceeds the indirect decrease in observed emissions resulting from decreased actual emissions due to an increase in monitoring equipment $(\partial a_i/\partial e_i)\ (\partial e_i/\partial K)$.

REFERENCES

Basar, T., Olsder, G.J. (1982). *Dynamic Non-cooperative Game Theory*, Academic Press, London (1982).

Brock, W.A., Malliaris, A.G. (1989). *Differential Equations, Stability and Chaos in Dynamic Economics*, North-Holland, Amsterdam.

Fershtman, C. (1987). "Identification of Classes of Differential Games for Which the Open Loop is a Degenerate Feedback Nash Equilibrium", *Journal of Optimal Theory Application*, 55(2), pp. 217–231.

Malik, A. (1990). "Markets for Pollution Control When Firms are Non-Compliant", *Journal of Environmental Economics and Management*, 18, pp. 97–106.

Meran, G., Schwalbe, U. (1987). "Pollution Control and Collective Penalties", *Journal of Institutional Theoretical Economics*, 143, pp. 616–629.

Oniki, H. (1973). "Comparative Dynamics (Sensitivity Analysis) in Optimal Control Theory", *Journal of Economic Theory*, 6, pp. 265–283.

Segerson, K. (1988). "Uncertainty and Incentives in Nonpoint Pollution Control", *Journal of Environmental Economics and Management*, 15, pp. 87–98.

Xepapadeas, A. (1991). "Environmental Policy under Imperfect Information: Incentives and Moral Hazard", *Journal of Environmental Economics and Management*, 20, pp. 113–126.

Xepapadeas, A. (1992). "Environmental Policy Design and Dynamic Nonpoint Source Pollution", *Journal of Environmental Economics and Management*, 23, pp. 22–39.

4. Nonpoint Tournaments

RAMU GOVINDASAMY, JOSEPH A. HERRIGES AND
JASON F. SHOGREN

Department of Economics, Iowa State University, Ames, IA, U.S.A

1. INTRODUCTION

In its final report to the U.S. Congress on section 319 of the Clean Water Act, the Environmental Protection Agency (EPA) states that "...information indicates very clearly that nonpoint source pollution has caused severe damage to aquatic communities nationwide and has destroyed the aesthetic values of many of our treasured recreational waters" [USEPA (1992, p. 1–2)]. The report contends that 16 percent (206,179 miles) of the rivers in 40 surveyed states and 20 percent (5.4 million surface acres) of the lakes have been impacted. The largest culprits remain nutrients, toxins, and siltation from agricultural and urban runoff. The report concludes by stressing the need to use the power of market incentives to efficiently prevent and control these nonpoint source (NPS) pollutants. By unleashing market forces, nonpoint pollution can be reduced in the most cost-effective way, thereby freeing resources to address other important problems.

Two market-incentive mechanisms have been recently proposed by economists to control NPS problems. Segerson (1988) argued for a tax on absolute levels of ambient concentration to be paid by the nonpoint polluter. The tax would charge the source for the marginal social damage of the ambient concentration level in surface and groundwater. But as Cabe and Herriges (1992) note the vast information requirements needed to implement the tax on absolute levels is likely to restrict its attractiveness. The tax requires collecting site-specific data on the complex environmental fate and transport system and the polluter's prior beliefs about this system. Without this detailed set of data, the tax will be misspecified and will not achieve its desired goal.

C. Dosi & T. Tomasi (eds.), Nonpoint Source Pollution Regulation: Issues and Analysis, 87–105.

At the other extreme, Xepapadeas (1991) has proposed a random penalty mechanism that requires a minimal amount of information to be implemented. If total ambient concentration exceeds a mandated level, the mechanism would randomly select one NPS and fine it. Site-specific monitoring of each NPS is unnecessary. In addition, given risk averse polluters, the random penalty mechanism is budget balancing and no external funds are required to implement the scheme. Despite its attractiveness in terms of efficiency, the political feasibility of a random penalty scheme remains a major question mark. The idea that one producer will be picked at random and punished, even if he or she is totally compliant, runs against the grain. Consequently, we are rather guarded as to the likelihood that a random penalty scheme will play a major role in NPS control.

This chapter offers a third mechanism: nonpoint tournaments. Nonpoint tournaments provide a middle ground between the ambient tax and random penalty schemes. Derived directly from the labor literature (Lazear and Rosen, 1981; Nalebuff and Stiglitz, 1983; Green and Stokey, 1983; Carmichael, 1983; O'Keeffe et al., 1984; Rosen, 1986; Malcomson, 1986; Bull et al., 1987; Drago and Heywood, 1989; Ehrenberg and Bognanno, 1990), a nonpoint tournament uses readily available information on input use or pollution abatement effort to construct a relative ranking of the NPS polluters. Each source is then rewarded or penalized based solely on the ranking of its pollution abatement effort relative to the abatement effort of all other sources. The advantage of the tournament approach is that ordinal information is typically less costly to obtain than cardinal information. The risk is that the information used to rank firms may be incomplete, providing an incorrect ranking of firms or an inappropriate signal regarding pollution abatement.

As a motivating example, consider agri-chemical contamination of surface and groundwater. Assume there are several farms irrigated by a reservoir. The reservoir is normally built in an elevated area and the water is drawn using gravitational force. The farms located downstream are irrigated through the channels from the reservoir. The transport of pesticides and fertilizers through irrigated water to the downstream farms is a common problem associated with agriculture. Downstream water contamination would reflect the level of pollution caused by the entire set of farms in the area. But it is extremely costly to attribute the absolute level of groundwater contamination to each and every farm.

Under such conditions, a nonpoint tournament scheme might be successfully applied. The regulator would monitor the groundwater contamination for the entire area, rank farms based on their pollution abatement effort or input use, and then penalize one or more of the lowest ranking farms if the ambient concentration for the entire area exceeds the optimal level. Alternatively, the regulator could reward only the highest ranking farms if the ambient concentration falls below the optimal level. This is a nonpoint tournament. Rewards or penalties depend on the relative rank of the farms, not on the absolute level of pollution emissions.

The chapter proceeds as follows. Section 2 sets forth the background of rank order tournaments in the labor literature, the basic analytical model, and the Pigouvian tax we will use as a comparative benchmark. Section 3 introduces the nonpoint tournament with a reward and penalty scheme. The tournament and tax are then compared in terms of their information requirements. Section 4 discusses three conditions under which nonpoint tournament will fail. Finally, we offer our conclusions in Section 5.

2. PRELIMINARIES, ANALYTICAL FRAMEWORK, AND BENCHMARK TAX

2.1. Preliminaries

The labor theory of rank order tournaments evolved to explain real world examples such as the competition for managerial posts, professional golf or tennis tournaments, horse races, and classroom grading. Since labor effort is generally unobservable or observed with error, the tournament ranks competitors based on observable output and then distributes rewards based on the final ranking. Since rankings are relative, the tournament requires less monitoring than traditional piece rate schemes which make compensation a function only of absolute output.

The seminal article by Lazear and Rosen (1981) develops the basic tournament structure. Therein, workers influence output by investing in unobservable effort (e.g., education, job training, etc.). The relationship between output and effort is assumed to be linear, with

$$q_j = \mu_j + \epsilon_j \tag{1}$$

where μ_j denotes the effort level (i.e., a measure of skill or average output) and ϵ_j denotes the realization of the random factor or shock with a mean of zero and a variance of σ^2. Effort is chosen by the worker prior to the realization of the random factor. The random factor can be thought of as the lifetime luck, such as an ability factor, which is revealed slowly over the worker's lifetime. The individual worker's cost of effort is given by $C(\mu)$ with $C' > 0$ and $C'' > 0$, where primes denote the relevant derivatives. The random term ϵ_j is assumed to be i.i.d. across individuals.

Lazear and Rosen (1981) begin by demonstrating that, with free entry and competition among workers, a piece rate system can be used to yield an efficient level of investment, with

$$C'(\mu) = V \tag{2}$$

where V denotes the value of unit output to the firm. That is, by setting the piece rate equal to V, the worker will equate their marginal cost of investment to the marginal social return from that effort, resulting in piece rates that are efficient.

Lazear and Rosen (1981) then show that a tournament between two players can be designed to achieve the same efficiency condition as (2). They consider a fixed prize (W_1) awarded to the winner of the tournament and a fixed prize (W_2) awarded to the loser. Winning is determined by the largest drawing of output, q. It is shown that the worker's effort increases as the spread between W_1 and W_2 increases. At the equilibrium, with a zero profit condition for the firms, they show that a rank order tournament yields the efficient level of labor effort. Lazear and Rosen conclude that in the case of risk neutral workers both tournaments and piece rate can be used to achieve an efficient allocation of resources.

Nalebuff and Stiglitz (1983), in extending the work of Lazear and Rosen, emphasize that rank order tournaments will be desirable when (a) the worker's input is not directly observable; (b) there is a stochastic relationship between the input and output; and (c) the stochastic disturbances which affect the relationship between input and output of different workers are correlated. Nalebuff and Stiglitz (1983) show that tournaments perform better than individualistic reward structures when there is a large source of uncertainty common among the agents. Green and Stokey (1983) also show that, if there is a large number of agents, a principal who cannot observe the common shock term can do as well as one who observes the common shock, by adopting a tournament compensation scheme. Finally, Malcomson (1986) generalized the equivalence of piece rates and tournaments under a broad set of conditions on preferences and technology.

2.2. *Analytical Framework*

In developing a nonpoint tournament model, we follow Lazear and Rosen's (1981) basic structure, with three changes. First, these authors assume that output is observable, but effort is unobservable. We assume the opposite, i.e., the overall level of pollution is unobservable, but pollution abatement effort (or an index of abatement effort consisting of proxy qualities or quantities) is observable. Second, they assume that output is linearly related to effort, with the ranking of the agents based on output. We assume effort is nonlinearly related to abatement and that the ranking of the firm depends upon their effort. Third, they use a fixed reward structure to induce optimal effort, whereas we also consider a flexible penalty scheme to induce socially optimal effort.

Consider an economy with two firms ($i = 1, 2$) that in the course of producing an output emit nonpoint pollution into the environment. Let Π_i^o represent the profit of the ith firm if they invest no effort towards controlling pollution. Let A_i represent the actual pollution abatement by firm i, while I_i represent the firm's observable effort towards controlling pollution. In the labor context a single valued measure of effort is considered. In the environmental arena, however, pollution abatement effort will typically be multidimensional, requiring the use of an effort index for the purposes of ranking. Consider the case of soil erosion control. An effort index (which can be viewed as the erosion control production function) could be assigned to each firm based on a vector of observable elements, such as the type of plow used for cultivation, the construction of contours, the use of strip cropping, and the method of irrigation, as well as the slope and rainfall in the area. The Universal Soil Loss Equation (USLE), for example, would provide a simple and readily available basis for initially quantifying the impact of each type of effort. More complex models, capturing the dynamics of soil erosion within a watershed, could be used to refine the effort index for erosion control. These models include the AGNSP model developed in Young et al., (1989) and the SEDEC model developed in Bouzaher et al., (1990).

The relation between A_i and I_i is assumed to take the form

$$A_i = f(I_i) + \epsilon_i, \tag{3}$$

where $f(I_i)$ is the non-linear transformation of effort to abatement, with $f'(I_i) > 0$, and ϵ_i is a random factor with $E(\epsilon_i) = 0$ and $\text{Var}(\epsilon_i) = \sigma_\epsilon^2$. In the context of soil erosion, ϵ_i would include the influence of weather on sedimentation flows from a particular farm. When there is heavy rain or wind, there will be increased runoff compared to a situation with no rainfall. The error term would also capture variations in the care and ability with which individual farmers carried out specific erosion abatement practices and unknown characteristics of the farm.

Finally, let $C(I_i)$ represent the cost of effort in pollution abatement with $C'(I_i) > 0, C''(I_i) > 0$, and $C(0) = 0$. The profit maximization problem of firm i is then given by

$$\max_{I_i} \pi_i = \Pi_i^o - C(I_i), \quad i = 1,2.$$

If there is no government intervention, a profit maximizing firm's optimal pollution abatement effort will be zero, since $C(I_i)$ negatively contributes toward profits.

2.3. *Comparative Benchmark : Pigouvian Tax*

Although a variety of mechanisms have been promoted to control nonpoint pollution, the Pigouvian tax is the classic. We now consider the use of the tax as a benchmark by which to evaluate the efficiency of the nonpoint tournament.

With no government intervention, a profit maximizing firm will set the effort level at zero, since the cost of effort negatively contributes towards the firm's profits. A Pigouvian tax can be designed to encourage the firm to choose the socially optimal effort level in pollution abatement. The Pigouvian tax in the nonpoint arena is similar to the piece rate approach to paying workers considered in Lazear and Rosen (1981). Let t be the per unit tax imposed by the regulator on both firms 1 and 2 for each unit of unabated pollution. With the Pigouvian tax, the firm's expected profits are given by

$$E\Pi_i = \Pi_i^o - t[X_i(0) - f(I_i) - C(I_i)], \quad i = 1,2. \tag{4}$$

where $X_i(0)$ represents the pollution contribution from the ith firm with no abatement effort and $[X_i(0) - f(I_i)]$ is the expected net contribution from the ith firm after abatement. The abatement level A_i is defined as the difference between ambient pollution contribution with no control, $X_i(0)$, and ambient pollution contribution with some investment toward controlling pollution, $X_i(I_i)$. That is, $A_i \equiv X_i(0) - X_i(I_i)$. The expected profits of the firm are the profits with no abatement minus the tax and abatement costs. Substituting equation (3) into equation (4), and using the fact that $E(A_i) = f(I_i)$ and $E(\epsilon_i) = 0$, the first-order conditions corresponding to the firm's profit maximization problem are

$$tf'(I_i) = C'(I_i), \quad i = 1,2 \tag{5}$$

i.e., the firm equates the marginal costs and benefits of effort.

Let λ represent the value to the society of a unit of pollution abatement. The objective of the social planner is to solve

$$\max_{I_i} E(M) = \{\Pi_1^o - C(I_1) + \Pi_2^o - C(I_2)\} + \lambda[f(I_1) + f(I_2)], \quad i = 1,2. \quad (6)$$

where $E(M)$ represents the expected welfare to the society as the sum of the firms' net profits and the welfare gains from pollution abatement. The corresponding first order conditions are given by

$$\lambda f'(I_i) = C'(I_i), \quad i = 1,2. \quad (7)$$

The social marginal benefit from pollution abatement effort equals the firms' marginal cost of that effort. The regulator can set t such that $t = \lambda$, so that the social and individual first order conditions coincide inducing optimal effort by firms – our comparative benchmark.

3. NONPOINT TOURNAMENTS

We now consider the nonpoint tournament. Both a reward and penalty scheme are examined. If we can show that the tournament induces the same efficient socially optimal effort as that of the Pigouvian tax, then given that the relative ranking of firms requires less information than assessing absolute levels of pollution abatement, a tournament will save resources relative to the Pigouvian tax.

3.1. Nonpoint Tournament With a Fixed Reward Scheme

Consider two firms polluting the environment, with the regulator setting up a rank order tournament between them in order to control pollution. Let the firms follow the pollution abatement effort as given by equation (3). Let P_W be the prize for the winner and P_L be the prize for the loser. The firms will be ranked on the basis of their pollution abatement effort with the firm with highest abatement effort chosen as the winner of the game. This is the nonpoint tournament, since winning depends on the relative rank and not on the absolute abatement level. Let $\hat{A}_i$ indicate the socially optimal level of abatement by firm i and $\hat{A} \equiv \hat{A}_1 + \hat{A}_2$ denote the optimal level of total abatement. For the purpose of budget balancing, the prizes P_W and P_L can be set such that

$$P_W + P_L = \lambda \hat{A} \quad (8)$$

where λ represents the value for unit abatement of pollution. Thus, one of the regulator's roles is to distribute to the abating firms the social gains from their abatement effort. Once the prize is set up, firms will increase expenditure on pollution abatement effort until the marginal cost of effort equals the marginal impact of that effort on their expected prize. The prizes P_W and P_L should be fixed such that the firms will choose the optimal abatement levels $\hat{A}_1$ and $\hat{A}_2$.

Consider the ith risk neutral firm's expected profits under this tournament scheme

$$\Pi_i^o + \beta[P_W - C(I_i)] + (1-\beta)[P_L - C(I_i)], \quad i = 1,2 \tag{9}$$

Equation (9) can be rewritten as

$$\Pi_i^o + [\beta P_W + (1-\beta)P_L] - C(I_i), \quad i = 1,2 \tag{10}$$

where β is the probability of winning the prize P_W,

$$\beta = Prob(A_1 > A_2), \tag{11}$$

and the term in square brackets in (10) denotes the firm's expected prize. Firms can influence the probability of winning the prize P_W by investing more in pollution control. But there is a trade-off between winning the prize P_W and expending effort in pollution abatement. Substituting equation (3) for A_i in equation (11) yields

$$\begin{aligned} \beta &= Prob([f(I_1) - f(I_2)] > \epsilon_2 - \epsilon_1) \\ &= Prob([f(I_1) - f(I_2)] > \omega) \\ &= G[f(I_1) - f(I_2)] \end{aligned} \tag{12}$$

where

$$\omega \equiv \epsilon_2 - \epsilon_1, \tag{13}$$

and G is the cumulative distribution function of ω, with $E(\omega) = 0$ and $\text{Var}(\omega) = 2\sigma_\epsilon^2 - 2Cov(\epsilon_1, \epsilon_2)$.

Each firm will choose I to maximize (9). For an interior solution, this implies

$$(P_W - P_L)(\partial\beta/\partial I_i) - C'(I_i) = 0, \quad i = 1,2 \tag{14}$$

and

$$(P_W - P_L)(\partial^2\beta/\partial I_i^2) - C''(I_i) < 0. \quad i = 1,2 \tag{15}$$

Equation (14) illustrates that the spread $(P_W - P_L)$ plays a key role in influencing the optimal level of effort. As the spread increases there is more incentive for the firms to invest in pollution control because the potential gains from winning increase.

Assuming Nash-Cournot play among the two players, firm 1 takes I_2 as given when determining its effort in pollution abatement and conversely for firm 2. Taking the derivative of equation (12) with respect to I_1 yields

$$\partial\beta/\partial I_1 = \partial G[f(I_1) - f(I_2)]/\partial I_1 = g[f(I_1) - f(I_2)]f'(I_1) > 0, \tag{16}$$

where $g(.)$ denotes the probability density function of ω. Substituting equation (16) into (14), firm 1's implicit reaction function is

$$g[f(I_1) - f(I_2)]f'(I_1)(P_W - P_L) - C'(I_1) = 0. \tag{17}$$

The slope of the explicit reaction function is

$$\frac{dI_1}{dI_2} = \frac{g'f'(I_2)f'(I_1)(P_L - P_W)}{g'f'(I_1)^2(P_L - P_W) + gf''(I_1)(P_L - P_W) - C''(I_1)}$$

This slope depends on the sign of g'. If $g' > 0$, then $(dI_1/dI_2) < 0$ and if $g' < 0$, then $(dI_1/dI_2) > 0$. The existence of the Nash equilibrium depends on the slopes of reaction functions of firms 1 and 2. See Cornes and Sandler (1986) for details.

Firm 2's reaction function is symmetric to (17). When a Nash equilibrium exists, symmetry implies that $I_1 = I_2 = I_N$ and $\beta = G(0) = 1/2$. That is, the tournament's outcome is random, with the probability of winning or losing set at 50 percent. At this Nash equilibrium, equation (17) becomes

$$C'(I_N) = (P_W - P_L)g(0)f'(I_N). \tag{18}$$

In order to establish the socially optimal level of abatement effort, the regulator must then set

$$(P_W - P_L) = \lambda/g(0). \tag{19}$$

In general, one would expect that as, the uncertainty regarding realized abatement increased (i.e., as Var(ϵ) increased), the spread between P_W and P_L would have to be increased. This is reflected in equation (19), since $(P_W - P_L)$ is inversely related to $g(0)$.

Substituting equation (19) into equation (18), the firm's profit maximization process will lead them to equate their marginal private cost $(C'(I_N))$ to the marginal social benefit $(\lambda f'(I_N))$ of their pollution abatement efforts – the exact condition derived from the benchmark Pigouvian taxation scheme.

The net social gains returns from optimal pollution abatement is $\lambda \hat{A}$, where the total prize offered to the two firms is $(P_W + P_L)$. For budget balancing, the total expected gains should equal total prize paid out,

$$\lambda \hat{A} = (P_W + P_L). \tag{20}$$

Substituting (20) in the firm's profit function, with $\beta = 0.5$, the firm's expected profit at the equilibrium effort strategy is

$$\Pi_i^o + \lambda \hat{A}/2 - C(I_N) \tag{21}$$

At the equilibrium, winning or losing the prize is purely random, since $\beta = 0.5$ and $I_1 = I_2 = I_N$. Holmström (1982) and Rasmusen (1987) indicate that budget balancing with risk neutrality leads to non-compliance. Budget balancing with risk neutrality works in tournaments due to nonlinearity introduced by the endogenous probability of winning the prize.

There are two characteristics of the above tournament structure that are worth noting at this point. First, the prizes (P_W and P_L) are fixed irrespective of the actions of the firms. One can combine equations (18) and (20) to solve explicitly for P_W and P_L, yielding:

$$P_W = \lambda[\hat{A} + 1/g(0)]/2 > 0 \tag{22}$$

and

$$P_L = \lambda[\hat{A} - 1/g(0)]/2. \tag{23}$$

While the loser's prize may become negative, the firm's expected prize $(\lambda \hat{A}/2)$ remains positive. The advantage of the fixed prize structure is that it commits the regulator to allocate to the contributing firms the total gains from its target level of pollution abatement. This avoids the potential moral hazard problem that the regulator might falsely claim a lower level of abatement and keep the surplus gains for itself. The disadvantage of this approach is that the potential for collusion among the regulated firms is enhanced. If the two firms successfully collude, the overall reward level remains unchanged, since it is independent of the actual level of pollution abatement.

Second, the reward scheme assigns to the firms the property rights to pollute with the total value of environmental quality improvements being distributed back to the firms. However, if the property right to pollute belongs to society, the firms should incur the cost of pollution, and the increased value of environmental quality due to abatement of pollution should be distributed back to society. Even though this scheme will be no longer be budget balancing, a penalty scheme can be administered to induce socially optimal effort levels among the firms.

3.2. Nonpoint Tournament With A Variable Penalty Scheme

In the penalty version of the nonpoint tournament, the firm's expected "prize" is negative when the actual abatement is less than the optimal abatement for the entire region. As before, it is assumed that the environmental authority is maximizing social welfare. However, now the social welfare gains due to abatement are redistributed back to society, not the firms. The prizes are set such that

$$P_W + P_L = \lambda[A^* - \hat{A}] \leq 0, \tag{24}$$

where $A^* = A_1^* + A_2^*$ represents the total actual abatement for the entire region and $\hat{A} = [\hat{A}_1 + \hat{A}_2]$ represents the optimal pollution abatement for the region set by the regulator. In addition, the individual penalties P_W and P_L are now specified as shares of the overall cost of suboptimal abatement. That is,

$$P_W = S_W[\lambda(A^* - \hat{A})] \tag{25}$$

and

$$P_L = S_L[\lambda(A^* - \hat{A})], \tag{26}$$

where S_W and S_L are fixed shares with $S_W + S_L = 1$ and $S_L > S_W$. The size of both prizes (P_L and P_W) are now variable, depending on the realized ambient concentration of pollution. As A^* moves away from $\hat{A}$, the prizes will become more negative. When $A^* = \hat{A}$, the right hand side of (24) becomes zero and a penalty is imposed on neither firm.

Using the penalty structure in equations (25) and (26), firm i's expected profits become,

$$\Pi_i^o + [\beta S_W + (1-\beta)S_L]\lambda[f(I_1) + f(I_2) - \hat{A}] - C(I_i), \quad i = 1,2 \tag{27}$$

As before, the corresponding first order conditions for firm 1's profit maximization yield its implicit reaction function to firm 2's effort:

$$\begin{aligned} &g[f(I_1) - f(I_2)]f'(I_1)(P_W - P_L) + \\ &[\beta S_W + (1-\beta)S_L]\lambda f'(I_1) - C'(I_1) = 0 \end{aligned} \tag{28}$$

The additional term in equation (28), when compared to equation (17), reflects the fact that a firm's individual effort influences not only the probability that it will win the tournament, but also the size of the "prize" to be won. Firm 2's reaction function is symmetric to (28). When a Nash equilibrium exists, symmetry again implies that $I_1 = I_2 = I_n$ and $\beta = G(0) = 1/2$. At the Nash equilibrium, equation (28) becomes:

$$\begin{aligned} C'(I_n) &= (P_W - P_L) \cdot g(0) f'(I_n) + [(S_W + S_L)/2] \lambda f'(I_n) \\ &= [(P_W - P_L) g(0) + (\lambda/2)] f'(I_n) \end{aligned} \qquad (29)$$

In order to induce the socially optimal level of abatement effort, the regulator would need to set P_W and P_L to satisfy

$$(P_W - P_L) = \lambda/[2g(0)]. \qquad (30)$$

Substituting Equation (30) into the reaction function in Equation (29) yields the optimality condition $\lambda f'(I_n) = C'(I)$. Comparing equations (30) and (19), we note that the variable nature of the penalty scheme allows the required spread between the winning and losing prizes to be halved.

The advantage of the variable penalty scheme lies in the reduced incentive for firms to collude. When the firms collude and total actual abatement is reduced, the penalty on both the firms is increased. As the total actual abatement approaches the optimal level, a firm's expected penalty will approach zero.

3.3. A Comparison Of Nonpoint Tournaments And Pigouvian Tax

We have shown that both the nonpoint tournament and Pigouvian tax can be designed so as to achieve the same efficiency conditions. However, the two mechanisms differ in terms of their information requirements and in terms of how common sources of error impact their efficacy.

First, consider information requirements. Pigouvian taxation requires information on either (1) the actual abatement level of each firm, A_i^* $(i = 1, 2)$, as well as the total ambient concentration, or (2) I_i, $f(I_i)$, and ϵ_i in order to accurately assess the pollution base. In addition, information on firm level abatement costs (i.e., the function $C(I_i)$) and marginal damage costs associated with the pollutant (λ) are needed to set the optimal tax rate. If we have information only I_i and $f(I_i)$, A_i cannot be determined without error. The information required for a tournament is more limited, since it depends upon the relative performance of each firm. Abatement cost structures and total abatement are still required in order to set the optimal prize levels. However, each firm does not need to be monitored for absolute actual pollution contributions.

Second, it is likely that error structure linking abatement effort (I_i) with abatement (A_i) will be correlated across individuals. This correlation in the error term, due to common weather factors, geological conditions, etc., affects Pigouvian taxes and rank order tournaments in different ways. Formally, consider dividing the error into two distinct terms: (1) common sources of error, η, and (2) a firm specific error, ρ_i, with $\epsilon_i = \eta + \rho_i, E(\eta) = E(\rho_i) = 0$, and $Cov(\eta, \rho_i) = 0$. Returning to the soil erosion example, as the farmer invests more effort in soil erosion control by contouring and strip cropping, better erosion control can be achieved. The term η would represent common weather factors, such as wind velocity, that affect the ability of all farmers to control soil erosion, whereas ρ_i would include variation in ability and conditions across firms. When a tournament is used, the regulator need not observe η since the tournament is a relative payment scheme. In the extreme, with $\rho_i \equiv 0 \forall i$, the rank of I_i will match the rank of A_i exactly, regardless of the level of η. The Pigouvian tax, however, would require information on η. A similar situation arises when there is a *common* source of measurement error in assessing the actual level of pollution abatement effort. The tournament mechanism is unaffected, while the Pigouvian tax will be incorrectly assessed.

4. CONDITIONS WHERE NONPOINT TOURNAMENTS MAY FAIL

Tournaments will fail if the regulator misranks the players, or if the players are induced to provide suboptimal level (or mix) of abatement effort. The labor literature has identified a number of potential problem areas. Drago and Turnbull (1988) identify three conditoins under which tournaments may fail: (1) when effort can be precisely inferred from output, (2) when sabotage arises among the competitors, and (3) when collusion develops among workers. Dye (1984) identifies four additional situations under which tournaments may fail: (1) if one of the agents is as twice productive as the other, but the principal believes them to be of identical ability; (2) if tournaments are subject to a form of instability induced by the introduction of other compensation schemes; (3) if there are few players, enhancing the possibility of collusion; and (4) if, in the case of single dimensional output, it does not make much of a difference between ranking the firms and monitoring the level of the output. In this section, we focus on three conditions which may lead to the failure of nonpoint tournaments: (1) errors in ranking due to the size and makeup of the error term, (2) collusion and sabotage by firms, and (3) heterogeneity among the participating firms.

4.1. *Mistakes Due To Error Term* (ϵ)

The curvature of $f(I)$ and uncertainty due to ϵ play a major role in the potential for error in ranking the pollution abatement achieved by individual firms. When the variance of ϵ is high, there is an increased probability that the rank of I_i will not match the rank of A_i. The ranking of firms is more sensitive to the error term when the curvature of $f(I)$ is flat. This section examines the potential for misranking and its relationship to the curvature of the transformation function and the variation in ϵ.

Suppose that we observe our two firms, 1 and 2, with $I_1 < I_2$. The probability (θ) that we will incorrectly rank firm 2 ahead of firm 1, based on their abatement effort, can be written as a function of $\Delta \equiv I_2 - I_1 > 0$ and the slope of the transformation function, f. Specifically,

$$\begin{aligned} \theta &\equiv Prob\{A_1 > A_2|\Delta\} \\ &= Prob\{f(I_1) - f(I_2) > \omega|\Delta\} \\ &= G[f(I_1) - f(I_2)] \\ &\approx G[-\Delta f'(I_2)] \end{aligned} \tag{31}$$

where $\omega \equiv \epsilon_2 - \epsilon_1$. Equation (31) indicates that the probability of incorrectly ranking the two firms increases (a) as the slope of f(I) diminishes and (b) as the difference between the effort levels of the two firms (Δ) shrinks.

Figure 1 illustrates the effects of both the curvature of the transformation function and the uncertainty captured by the error term ϵ. Consider first the steep transformation function, $\bar{f}(I)$. If there are no random effects ($\epsilon = 0$), then firm 1 and 2's investment in effort levels I_1 and I_2 will yield abatement levels $\bar{A}_1$ and $\bar{A}_2$, respectively. The ranking is consistent – firm 2 uses more input ($I_2 > I_1$) and is ranked higher than firm 1, $\bar{A}_2 > \bar{A}_1$. If there is a common error such that $\epsilon_1 = \epsilon_2 = +\epsilon$ or $\epsilon_1 = \epsilon_2 = -\epsilon$, the ranking remains consistent. For example, $\epsilon_1 = \epsilon_2 = +\epsilon$, then $I_2 > I_1$ still implies $\bar{A}_2 > \bar{A}_1$. This is illustrated by points a and b in Figure 1. Ranking may be preserved even if firms have different error terms, as in Figure 1 when $\epsilon_1 = 0$ and $\epsilon_2 = -\epsilon$, with points *c* and *d* showing that the correct ranking is preserved. However, the potential for misranking increases as the transformation function becomes flatter. If the transformation function is

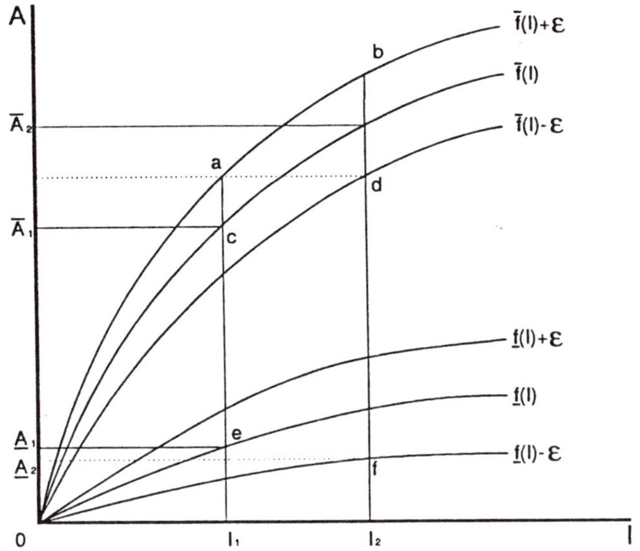

Fig. 1. Mistakes Due to Random Effects(ϵ)

given by $\underline{f}(I)$ in Figure 1, and we again have $\epsilon_1 = 0$ and $\epsilon_2 = -\epsilon$, we will misrank the firms. Points e and f illustrate that for inputs $I_1 < I_2$, the ranking for abatement is $\underline{A}_1 > \underline{A}_2$.

Finally, Figure 2 shows that errors in ranking can occur even with relatively steep transformation functions if the size of the error is large. With $\epsilon_1 = 0$ and $\epsilon_2 = -\epsilon$, points a and b illustrate that correct ranking occurs, with $I_1 < I_2$ implying $A_1 < A_2$, the correct ranking of firms. However, if ϵ_2 increases to $-\hat{\epsilon}$, points a and c imply $\hat{A}_1 > \hat{A}_2$, and an error in abatement ranking will occur.

The successful use of nonpoint tournaments requires an appreciation of the nature of the transformation function and the error structure. This may force the regulator to expend resources to improve his or her understanding of the relationship between input and abatement and the error. At the extreme, the information requirements may approach the level needed for a Pigouvian tax, thereby decreasing the attractiveness of the tournament.

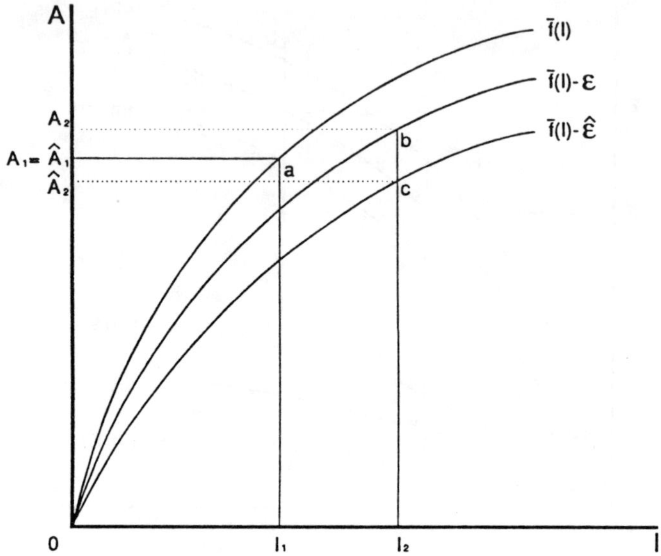

Fig. 2. Mistakes Due to the size of the Error Term

4.2. *Collusion And Sabotage Among Firms*

The competition provided by a nonpoint tournament induces the optimal abatement effort by firms. But competition exists only if there is no collusion among the firms. Since winning or losing depends on the relative rank, the firms have incentives to collude. If firms collude, a ranking might not lead to an efficient allocation of effort in pollution abatement. The potential for collusion increases with fewer numbers of firms. Figure 3 illustrates the problem of collusion. When the effort levels equal I_1 and I_2 for the two firms, the abatement will be at A_1 and A_2 under no collusion. If the firms collude, the ranking of the firm 1 and firm 2 could be maintained, but the effort levels could shift from (I_1, I_2) to (I_1', I_2'). This decreased effort would lead to a lower level of actual abatement (A_1' and A_2'). In the case of the fixed reward scheme, the firms would receive the same prices with reduced and suboptimal effort.

The variable penalty reduces collusion problems by tying the average prize to the total abatement effort. If the firms collude to produce less total abatement, the sum of penalties will go up. The firms as a group pay the sum of the pollution damages imposed on society, reducing the incentive to collude.

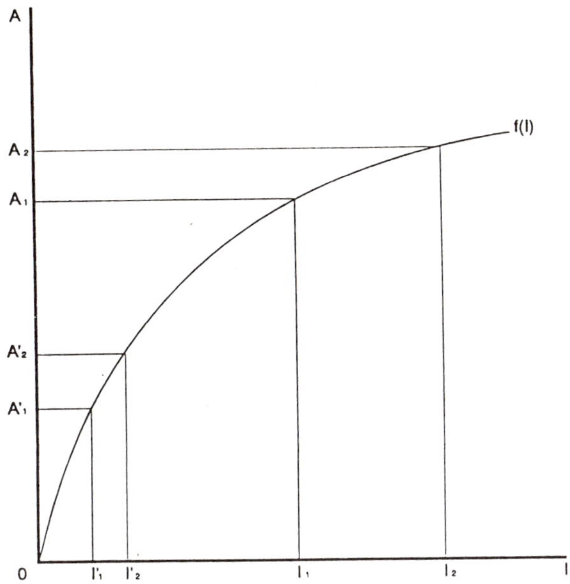

Fig. 3. Collusion or Sabotage

Collusion causes suboptimal effort at one extreme while sabotage causes suboptimal effort at the other. In case of two firms, when one sabotages the abatement by the other firm, the saboteur can win the prize with a low level of abatement effort. Again, the variable penalty scheme will reduce the incentive for sabotage by making the saboteur pay a fraction (S_L) of the social damages stemming from (a) their reduced abatement effort and (b) the reduced abatement by the sabotaged firm.

4.3. Heterogenous Firms

A major limitation associated with the application of tournaments to NPS pollution is the presence of heterogeneity among the contestants. In the example of groundwater contamination each producer can differ from others in characteristics such as weather, soil conditions, proximity to major aquifers, slope of the land, and water availability.

Firms may also differ in terms of their abatement costs. If a tournament is held among contestants of unequal abilities, Lazear and Rosen (1981) show that these two firms (one with high ability to invest and the other with low ability to invest) will not sort into their own ability level groups for competition. Rather both the firms will try to compete with the high ability firms because the rewards increase with the ability of the firms. As a result, competition between a high ability firm and a low ability firm (i.e., mixed contestants) leads to inefficient level of investment by the firms. However, if the heterogeneous nature of the firms are known to everyone, a handicap system can be adopted to elicit socially optimal efforts using tournaments.

5. CONCLUDING REMARKS

The nonpoint tournament provides a middle ground between the Pigouvian tax and random penalty scheme. First, one can overcome the problems associated with the random penalty mechanism such as penalizing firms that contribute less toward groundwater pollution by adopting a tournament, because the ranking should not penalize the least polluting firm. Second, the problems associated with perfect monitoring as in taxation can be minimized by using limited information and proxies to rank the firms. Third, a tournament does not require information on the common disturbances such as weather effects on the groundwater contamination by ranking of the firms. That is, by ranking two firms with a common weather factor, a regulator who cannot observe the common shock can do as well as one who observes the common shock. Therefore, if a Pigouvian tax can be administered, a tournament can be designed to attain the same allocation of resources in case of risk neutral firms using limited information and proxy variables. The use of limited information in a tournament leads to cost savings in acquiring information.

REFERENCES

Bouzaher, A., Braden, J., Johnson, G. (1990). "A Dynamic Programming Approach to a Class of Nonpoint Source Pollution Control Problems", *Management Science*, 36, pp. 1–15.

Bull, C., Schotter, A., Weigelt, K. (1987). "Tournaments and Piece Rates: An Experimental Study", *Journal of Political Economy*, 95, pp. 1–33.

Cabe, R., Herriges, J.A. (1992). "The Regulation of Non-Point Source Pollution Under Imperfect and Asymmetric Information", *Journal of Environmental Economics and Management*, 22, pp. 134–146.

Carmichael, L.H. (1983). "The Agent-Agents Problem: Payment by Relative Output",

Journal of Labor Economics, 1, pp. 50–65.

Cornes, R., Sandler, T. (1986). *The Theory of Externalities, Public Goods, and Club Goods*, Cambridge University Press, Cambridge.

Drago, R., Heywood, J.S. (1989). "Tournaments, Piece Rates, and the Shape of the Payoff Function", *Journal of Political Economy*, 97, pp. 992–998.

Drago, R., Turnbull, J.K. (1988). "The Incentive Effects of Tournaments with Positive Externalities Among Workers", *Southern Economic Journal*, 55, pp. 100–106.

Dye, R.A. (1984). "The Trouble with Tournaments", *Economic Inquiry*, 22, pp. 147–149.

Ehrenberg, R.G., Bognanno, M.L. (1990). "Do Tournaments Have Incentive Effects?", *Journal of Political Economy*, 98, pp. 1307–1324.

Green, J.R., Stokey, N.L. (1983). "A Comparison of Tournaments and Contracts", *Journal of Political Economy*, 91, pp. 349–364.

Holmstrom, B. (1982). "Moral Hazard in Teams", *Bell Journal of Economics*, 13, pp. 324–340.

Lazear, E.P., Rosen, S. (1981). "Rank-Order Tournaments as Optimum Labor Contracts", *Journal of Political Economy*, 8, pp. 841–864.

Malcomson, J.M. (1986). "Rank Order Contracts for a Principal with Many Agents", *Review of Economic Studies*, 53, pp. 807–817.

Nalebuff, B.J., Stiglitz, J.E. (1983). "Prizes and Incentives: Towards a General Theory of Compensation and Competition", *The Bell Journal of Economics*, 14, pp. 21–43.

O'Keeffe, M.W., Viscusi, K., Zeckhauser, R.J. (1984). "Economic Contests: Comparative Reward Schemes", *Journal of Labor Economics*, 2, pp. 27–56.

Rasmusen, E. (1987). "Moral Hazard in Risk-Averse Teams", *RAND Journal of Economics*, 18, pp. 428–435.

Rosen, S. (1986). "Prizes and Incentives in Elimination Tournaments", *American Economic Review*, 76, pp. 701–715.

Segerson, K. (1988). "Uncertainty and Incentives for Non-Point Source Pollution Control" *Journal of Environmental Economics and Management*, 15, pp. 87–98.

Stevens, S.S. (1968). "Measurement, Statistics, and the Schemapiric View", *Science*, 161, pp. 849–56.

United States Environmental Protection Agency (1992). *Managing Nonpoint Source Pollution*, Office of Water (WH-553), EPA-506/9–90, Washington, D.C.

Xepapadeas, A.P. (1991). "Environmental Policy Under Imperfect Information: Incentives and Moral Hazard", *Journal of Environmental Economics and Management*, 20, pp. 113–126.

Young, R.A., Onstad, C.A., Bosch, D.D., Anderson, W.P. (1989). "AGNPS: A-Nonpoint-Source Pollution Model for Evaluating Agricultural Watersheds", *Journal of Soil and Water Conservation*, Mar-Apr, pp. 168–173.

5. Nonpoint Source Externalities and Polluter's Site Quality Standards Under Incomplete Information

CESARE DOSI AND MICHELE MORETTO

Department of Economics, University of Padova, Italy

1. INTRODUCTION

There is basic consensus in the technical literature on the factors that distinguish nonpoint source (NPS) pollution (NPSP) from traditional point-source problems. These features may be summarised as follows.

Firstly, there are no simple add-on devices for intercepting and neutralising pollutant emissions so as to avoid or reduce discharges at the final receptor point(s).

Secondly, emissions at source cannot be generally monitored at reasonable costs, or inferred from observation of ambient pollutant concentrations. Depending on the problem at hand, this may be due to the number and dispersion of activities involved, and/or to the difficulty of correctly evaluating the delivery ratio – i.e. the ratio between ambient concentrations and pollutant emissions at source – because of the mode of conveyance and the often complex environmental fate of NPS pollutants.

Finally, besides production (or consumption) processes, the volume and quality of emissions and, consequently, the required abatement measures, are often significantly correlated to the physical features of the site where (suspected) polluting activities take place. For example, in cases where the same amount of a potentially polluting input is used, discharges may vary from firm to firm, according to the physical characteristics of their site. More importantly, what characterizes many NPSP problems – compared to traditional point-source phe-

C. Dosi & T. Tomasi (eds.), Nonpoint Source Pollution Regulation: Issues and Analysis, 107–136.

nomena – is that those features which affect the extent of pollutant emissions (e.g. the fraction of polluting inputs dispersed in the environment) are more susceptible to unforseeble exogeneous "shocks" which reduce the agent's control over the evolutionary pattern of his/her "type", and, consequently, over his/her discharges.

The first of the above mentioned NPSP features calls for regulatory policies focusing on the prevention of emission at source. This partly contrasts with the first generation of environmental policies which, particularly in water quality management, have mainly focused on technological devices that could intercept pollutant flows.

However, adoption of a preventive approach is undoubtedly complicated by the second of the above mentioned issues: i.e., the difficulty or even impossibility of monitoring individual emissions. This explains why the economic literature dealing with NPSP control has largely focused on the search for alternatives to "traditional" policy instruments based on observable emission at source, such as effluent taxes or emission standards. Two lines of research emerge in particular from this literature.

The first of these focuses on incentive mechanisms based on measurable ambient pollutant levels. Such incentives mainly take the form of tax/subsidy schemes based on deviations between observed and desired pollutant concentration levels at the receptor point (Segerson, 1988; Xepapadeas, 1992).

The second proposal, which we have referred elsewhere to as an "indirect approach" (Dosi and Moretto, 1993) suggests basing regulatory policies on estimated NPS emissions as determined by already available biophysical models (Griffin and Bromley, 1982; Shortle and Dunn, 1986; Dosi and Moretto, 1992). According to the proponents, these models, whose development was stimulated by the difficulty of assessing the extent of NPS emissions through comprehensive and large scale direct monitoring, may also provide a basis for designing regulatory instruments – such as economic incentives or management practice standards – provided the models are granted with "regulatory legitimacy" (Dosi and Moretto, 1992).

While granting such legitimacy would make the polluter's private information about the volume of effluents irrelevant[1], the regulator may still face problems of incomplete information and observation about the parameters required to feed the model itself. These parameters typically include production patterns (e.g. crop patterns) and management practices (e.g. amount and mode of utilization of potentially NPS water polluting inputs like chemical fertilizers and pesticides). Moreover, most of the available models explicitly account for the third of the above mentioned NPSP features, i.e. the role played by the physical characteristics of the site where suspected polluting activities take place in determining the extent of pollutant run offs, so that, other things being equal, estimated emissions and, consequently, the required abatement measures (e.g. changes in the amount and mode of application of nitrogen fertilizers) vary across firms according to the physical endowments of their site (e.g. water retention capacity of the soil).

The relative appeal of the two policy approaches depends on the particular situation at hand as well as on value judgements with regard to their economic and political viability. Besides other considerations, the first approach at first glance seems to be more promising whenever it is legitimate to believe that suspected polluters possess better information about the "technology" of ambient concentrations of pollutant abatement. On the other hand, the indirect approach seems to be more appropriate in those instances where polluters do not already possess or may easily acquire, at reasonable costs, information regarding pollutant release and transport processes (Dosi and Moretto, 1993)

Following a previous line of research (Dosi and Moretto, 1992; 1993), we concentrate in this paper on the indirect approach, focusing on physical characteristics of the suspected polluters' site ("site quality") which, according to the selected biophysical model, are considered key parameters to estimate (unobservable) individual emissions.

Special attention is devoted to the dynamics of the site quality. As in Dosi and Moretto (1993), the state variable site quality is assumed to be altered by firms' management practices ("current maintenance"), but its evolutionary pattern is also affected by stochastic exogeneous shocks in such a way that future site quality is uncertain with a variance growing along with the time horizon.

Moreover, we now consider the case where firms have access to a technology which allows one to regulate site quality instantaneously through a lump sum maintenance expenditure which applies only when the state variable hits a predetermined lower barrier.

Assuming that, besides influencing the extent of pollutant emissions, site quality also affects production possibilities (e.g. by altering the marginal productivity of a variable input), we first consider the firms' optimal barrier policy without public intervention, i.e. when no public control is exercised over the site quality admissability range and the firms bear the full cost of keeping their site from going below the desired minimum quality level[2]. Within our framework, the term "barrier policy" refers to the decision as to what is the minimum quality level at which lump sum maintenance expenditures come into effect.

We then turn to the environmental authority's problem of identifying what site quality range is socially desirable. We conduct the analysis under the assumption that the agency is not only unable to continuously observe or infer at reasonable costs the evolution of a site quality, but also cannot credibly commit itself to randomly monitoring the firms' compliance with the prescribed barrier policy.

We then consider the following scenario. Instead of searching for an optimal auditing policy – comprising the probability that an audit will be carried out and penalty for non-compliance with minimum quality standards – the agency offers the firms a subsidy, proportional to cost afforded whenever their site quality hits the lower barrier, so as to induce the firms to choose a site quality admissability range that is different from the one they would have chosen when bearing the full upkeeping costs. The firms will be given the subsidy every time they announce that the site quality has hit the lower barrier indicated by the agency which reserves the right to check the claim and that the firm has taken the steps necessary to push site quality upward.

Identifying the optimal agency barrier policy, comprising minimum site quality standards and subsidies for the firms' lump sum maintenance outlay, are the main topics of this paper.

The paper is organized as follows. In the next Section we provide the basic model and notation. Section 3 is devoted to solving the firm's dynamic optimization problem without public intervention; in the same Section we also explore how the firm's optimal barrier policy is affected by uncertainty on the evolution of the site quality. In Section 4 we turn to the agency's problem in order to identify the contract to be offered to the firm. It will be shown that the general features of such a contract are crucially dependent on the assumed technical relationship between pollutant discharges and the index reflecting the firm's site quality status, on the uncertainty with regard to its evolutionary pattern, as well as the social opportunity cost of public funds. Section 5 summarizes the results.

2. THE BASIC MODEL

2.1. *Site Quality, the Firm's Marketable Output, and Estimated Environmental Damage*

We consider a risk-neutral utilitarian agency in charge of regulating several competitive firms which have been identified as potential contributors to pollutant concentrations observed in one or more environmental media. The agency's welfare value depends on the firms' profits and instantaneous environmental damages over the planning horizon $[0, \infty)$.

Environmental damages are modeled as being dependent on the flows of pollutant emissions at source which, although not directly observable, can be estimated through available biophysical models.

According to the model which has been "politically legitimated", emissions are a function of the physical characteristics of a firm's location, $P_t^i = P(\theta_t^i)$, where θ_t^i represents an index for the i-th firm site quality at time t.

As well as affecting the amount of (estimated) pollutant emissions, the site quality also affects firms' production possibilities, i.e.

$$Q_t^i = Q(\theta_t^i) \tag{1}$$

where Q_t^i represents the i-th firm's marketable output at time t.

Assumption 1. Indicating with $d\theta_t > 0$ an improvement in the site overall quality, $Q(\theta)$ holds the following properties: $Q'(\theta) > 0$; $Q''(\theta) < 0$; $Q(0) = 0$; $Q'(0) = \infty$ and $Q'(\infty) = 0$.

For the sake of simplicity, we will consider the case where pollution damages are additively separable, so that under our assumptions regulating each firm individually is optimal. The ultimate harmful environmental impact associated with a firm's estimated pollutant emissions at time t are evaluated according to the following damage function:

$$D_t = D(\theta_t), \qquad D' \gtrless 0 \tag{2}$$

where the superscript i is dropped for convenience. Notice that the firm's estimated emissions and, consequently, instantaneous damages, can be either positively or negatively correlated with changes in the site quality. In fact, site quality changes which are expected to improve the firm's productive performance do not necessarily imply an improvement in the firm's "environmental" performance, i.e. a reduction in the amount of (estimated) pollutant discharges. For example, consider the case where emissions are simply represented by the amount of a polluting input, x, which is not completely "absorbed" during the production process and is then dispersed in the environment, and suppose that an improvement in the site quality allows the firm to shift upward the level of x beyond which its marginal productivity becomes negative. Even if emissions per unit of x were negatively correlated with site quality improvements, the latter would induce greater input consumption that might more than compensate for the fall in x dispersion per unit (see Dosi and Moretto, 1993).

2.2. *The Dynamics of the Site Quality*

The state variable θ_t follows an Itô's process, regulated with a lower barrier $S \geq 0$. At $\theta_t = S$, an infinitesimal control $dL_t$is applied to θ and gives it a "push" upward.

Overall, the stochastic differential equation for θ is

$$d\theta_t = [f(m_t, \theta_t) - \delta]\, \theta_t dt + \sigma \theta_t dz_t + dL_t \qquad (3)$$

where $f(.)$ describes the effect of the site's current maintenance expenditure (m), δ is the site quality constant depreciation rate, dz_t is the increment of a Wiener process or Brownian motion, and $\{L_t\}$ is a stochastic process which is right-continuous, non-negative and non-decreasing, and increases only when $\theta_t = S$ (Harrison, 1985, p.23).

Assumption 2. The function $f(.)$ describing the effect of current maintenance expenditure holds the following properties:

$$f_m > 0;\ f_{mm} < 0;\ f_\theta < 0;\ f(0, \theta) = 0$$

Control dL_t is applied only when θ hits S, so that a minimum degree of regulation is exercised to keep the state variable from going under the range $S \leq \theta_t < \infty$. The lower regulator dL_t is operated at a cost dC_t. This cost is a right-continuous, non-negative and non-decreasing process, and C_t increases only when L_t does. The increment of C_t is given by a cost function that we assume, for simplicity, to be linear.

Assumption 3. $dC_t = C(dL_t) = bdL_t, b > 0$ where b is a parameter constant over time.

Notice that while current maintenance m_t represents a variable cost, within our framework the maintenance expenditure dC_t takes on the role of a sunk investment cost.

2.3. *The Firm's Problem Without Public Intervention*

As is customary, it is assumed that the firm wishes to maximize its market value, i.e. its expected discounted cash flow over the planning horizon $[0, \infty)$.

According to (1) and (3) and setting the marketable output price equal to one, the firm's market value function is described by:

$$F(\theta_0; S) = E_0 \left\{ \int_0^{\infty} e^{-rt} \left[(Q(\theta_t) - m_t)\, dt - b dL_t \right] \mid \theta(0) = \theta_0 \right\} \quad (4)$$

$$S \leq \theta_t < \infty$$

where r is a constant discount rate.

The firm's problem is to choose a pattern of current maintenance expenditures $\{m_t\}$ which maximizes, for any barrier S, the value of (4) under the constraint described by equation (3), and then optimize with respect to S.

2.4. *The Agency's Problem*

If the agency were able to monitor the firm, it would mandate a pattern of current maintenance expenditures and a lower site quality barrier S, which differs from the choice the firm would make. In particular, as far as the barrier S is concerned, if the instantaneous damage function (2) is increasing in θ, the agency would impose a lower optimal barrier S than that chosen by the firm; if instantaneous damages decrease with θ, the reverse applies.

However, although the agency shares the same information as the firm on the production function (1) and the process governing the evolution of θ given by (3), we assume that it is neither able to mandate and control the firm's management practices, or to continuously monitor or to infer the realisations of process $\{\theta_t\}$.

We then consider the following policy. The agency foregoes any attempt to influence current maintenance expenditure decisions, and offers the firm a subsidy, proportional to the upkeeping cost required to regulate θ instantaneously, so as to induce the firm to choose a barrier S which is different from the one it would have otherwise chosen if it had had to bear the full cost of regulation. Accordingly, the agency's welfare function is:

$$W(\theta_0; S, \alpha) = E_0 \left\{ \int_0^\infty e^{-rt} \left[(Q(\theta_t) - m_t) \, dt - (1-\alpha) b dL_t \right] + \right.$$

$$\left. - \int_0^\infty e^{-rt} D(\theta_t) dt - (1+\rho)\alpha b \int_0^\infty e^{-rt} dL_t \mid \theta(0) = \theta_0 \right\} S \leq \theta_t < \infty \tag{5}$$

where $\alpha b (\alpha \geq 0)$ is the monetary transfer granted to the firm every time it claims to have touched the lower barrier S indicated by the agency, and $(1+\rho) > 1$ is the social opportunity cost of public funds.

The function (5) can be rewritten as follows

$$W(\theta_0; S, \alpha) = E_0 \left\{ \int_0^\infty e^{-rt} \left[(Q(\theta_t) - m_t) \, dt - b dL_t \right] + \right.$$

$$\left. - \int_0^\infty e^{-rt} D(\theta_t) dt - \rho \alpha b \int_0^\infty e^{-rt} dL_t \mid \theta(0) = \theta_0 \right\} \qquad S \leq \theta_t < \infty \tag{5'}$$

i.e. the agency's welfare value can be expressed as the sum of the market value of a firm bearing the full cost of regulating θ, net of the expected discounted flow of instantaneous damages and the expected discounted social cost of subsidizing the firm.

If we take S^* to indicate the optimal lower barrier which the firm would choose outside the relationship with the agency, the problem facing the latter may be formalized as follows:

$$\max_{s,\alpha} W(\theta_0; S, \alpha) \tag{6}$$

$$\text{s.t. } \{m_t\} = \arg\max \{F(\theta_0; S, \alpha) \text{ for given } S \text{ and } \alpha \},$$

$$F(\theta_0; S^*) \leq F(\theta_0; S, \alpha)$$

The first constraint reflects the hypothesis that the agency is unable to mandate and control the firm's current maintenance expenditure pattern. The second constraint reflects the requirement that ensures the firm's participation, i.e. acceptance of the contract (S, α). The agency must ensure that the firm has a market value greater than that which it could obtain outside the relationship with the agency.

3. THE FIRM'S DYNAMIC OPTIMIZATION PROBLEM WITHOUT PUBLIC INTERVENTION

3.1. Problem Set-up

In order to solve the program (6), we need to characterize the two constraints, i.e. to solve the firm's problem. To keep it mathematically tractable, we specify the production function (1) and the maintenance technology function, $f(.)$.

Assumption 4. $Q(\theta_t) = q\theta_t^{\nu}, \quad 0 < \nu < 1$ and $q > 0$;
$f(m_t, \theta_t) = m_t^{\xi}\theta_t^{-\gamma} \quad 0 < \xi < 1$ and $\gamma > 0$.

According to assumption 3 and 4, the firm's market value function (3) becomes

$$F(\theta_0; S) = E_0\left\{\left[\int_0^{\infty} e^{-rt}\left(q\theta_t^{\nu} - m_t\right)dt - bdL_t\right] \mid \theta(0) = \theta_0\right\} S \leq \theta_t < \infty \tag{7}$$

The firm maximizes (7) subject to equation (3), the constraint $m_t \geq 0$, and θ_0 is given. The sample path of $\{z_t\}$ is assumed to contain all the information relevant to the firm's problem, and $E_0\{.\}$ denotes conditional expectation taken at time zero over the distribution of $\{z_t\}$ and $\{\theta_t\}$ processes. While $\{z_t\}$ is exogeneous to the firm's problem, $\{\theta_t\}$ is determined by the optimal current maintenance expenditure pattern.

Since the choice concerning this pattern is independent from the control dL_t, the aim is to evaluate (7) for given S, choosing the optimal path $\{m_t\}$, and then identifying the lower barrier S^* which maximizes the firm's market value.

3.2. The Firm's Market Value for any Control S

Provided the optimal maintenance policy $m_t = m(\theta_t)$ is well defined, the following proposition holds.

Proposition 1 *Within the zone of no intervention ($S < \theta_t < \infty$), the process $\{\theta_t\}$ moves onward on its own, and the expected change in the firm's value function $F(.)$ – brought about by the pay-off flow $[q\theta_t^{\nu} - m(\theta_t)]$ and the effect of discounting – satisfies the following differential equation:*

$$\frac{1}{2}\sigma^2\theta_t^2 F_{\theta\theta} + \left(m(\theta_t)^{\xi}\theta_t^{-\gamma} - \delta\right)\theta_t F_{\theta} - rF + q\theta_t^{\nu} - m(\theta_t) = 0 \tag{8}$$

where $F_{\theta\theta}$ and F_{θ} are partial derivatives with respect to θ.

Proof: see Appendix A.

Equation (8) represents a linear differential equation in F, so that its general solution may be expressed as the sum of two parts: the general solution of the homogeneous equation (complementary function) and any particular solution,

$$F(\theta_0; S, h_1, h_2) = V(\theta_0) + h_1(S)F^1(\theta_0) + h_2(S)F^2(\theta_0) \qquad (9)$$

where $V(\theta_0)$ stands for a particular solution of (8), whilst F^1 and F^2 represent two solutions of the associated homogeneous equation and h_1 and h_2 are two constants to be determined.

The Particular Solution

A very convenient particular solution of (8) is the expected discounted flow given by (7), ignoring the barrier S and the cost of regulation:

$$V(\theta_0) = E_0\left\{\int_0^\infty e^{-rt}\left(q\theta_t^\nu - m_t\right)dt \mid \theta(0) = \theta_0\right\} \qquad 0 < \theta_t < \infty \quad (10)$$

If V is differentiable, (10) has to be the solution of the following dynamic programming equation:

$$rV = \max_m\left\{q\theta_t^\nu - m_t + \left(m_t^\xi\theta_t^{-\gamma} - \delta\right)\theta_t V_\theta + \frac{1}{2}\sigma^2\theta_t^2 V_{\theta\theta}\right\} \qquad (11)$$

By differentiating the right-hand-side of (11) with respect to m, we obtain the first order condition for optimality of the firm's current maintenance expenditure pattern, from which we obtain the following expression:

$$m_t = \left(\xi V_\theta \theta_t^{1-\alpha}\right)^{1/(1-\xi)} \qquad (12)$$

Equations (11) and (12) together can be expressed as a non-linear second order differential equation of parabolic type in V. As pointed out by Vorst (1987), Moretto (1991) and Dosi-Moretto (1993), such a differential equation can be explicitly solved by imposing some restrictions on the coefficients of the production and maintenance technology functions.

Assumption 5. $\xi = \frac{1}{2}$ and $\gamma = \frac{1}{2}\nu$

Notice that $\xi = 1/2$ is right in the middle of the domain of ξ, and by this restriction we find that the Hamilton-Jacobi-Bellman equation (11) has only linear terms in V_θ. The second restriction is for technical reasons.

Under assumption 5, the next proposition can be proved.

Proposition 2 *The solution of the dynamic programming equation (11) is:*

$$V(\theta_t) = M\theta_t^\nu \tag{13}$$

where:

$$\begin{aligned} M &= \frac{B - \sqrt{B^2 - 4Aq}}{2A} \\ A &= \frac{1}{4}\nu^2 > 0 \\ B &= r + \delta\nu - \frac{1}{2}\nu(\nu - 1)\sigma^2 > 0 \end{aligned}$$

For the proof of this proposition the reader is referred to Dosi and Moretto (1993).

From (13) the optimal current maintenance policy rule can be derived:[3]

$$m_t^* = \frac{1}{4}\nu^2 M^2 \theta_t^\nu \tag{14}$$

i.e. optimal maintenance expenditure is a smooth, monotone function of the current realization of the state variable θ_t.[4] Accordingly, the stochastic differential equation (3) reduces to the increment of a geometric Brownian motion with instantaneous regulation

$$d\theta_t = \mu\theta_t dt + \sigma\theta_t dz_t + dL_t \tag{15}$$

where $\mu = \frac{1}{2}\nu M - \delta$.

In fact, if we define the regulated process $\{\theta_t\}$ by the relationship $\theta_t = X_t l_t$ (where $\{X_t\}$ is a geometric Brownian motion $dX_t = \mu X_t dt + \sigma X_t dz_t$ and initial value $S \leq \theta_0 < \infty$), according to Harrison (1985, p. 22), we can identify $\{l_t\}$ as a process defined as $l_t = \inf\{S/X_\tau\}$, with $0 \leq \tau < t$, and $l_0 = 1$, which increases only when $\theta_t = S$. By applying the Itô's Lemma to θ_t, we get $d\theta_t = l_t dX_t + X_t dl_t = \mu X_t l_\tau dt + \sigma X_t l_t dz_t + X_t l_t (dl_t/l_t)$, which, assuming $dL_t = \theta_t (dl_t/l_t)$, gives (15).

The Complementary Function Solution

Omitting $[q\theta_t^{\nu} - m(\theta_t)]$ from (8) and substituting (14) for $m(\theta_t)$, we obtain the associated homogeneous function, which appears to be a differential equation of Euler type. To solve it, we can guess a functional form of type θ_t^K, and check by substitution if it works. It is easy to show that the solution is:

$$F^0(\theta_t; S, h_1, h_2) = h_1(S)\theta_t^{K1} + h_2(S)\theta_t^{K2} \tag{16}$$

where $K1$ and $K2$ are respectively the positive and negative roots of the quadratic equation

$$\frac{1}{2}\sigma^2 K^2 + \left(\mu - \frac{1}{2}\sigma^2\right) K - r = 0 \tag{17}$$

with

$$K1 = -\frac{1}{\sigma^2}\left[\left(\mu - \frac{1}{2}\sigma^2\right) - \sqrt{\left(\mu - \frac{1}{2}\sigma^2\right)^2 + 2r\sigma^2}\,\right] > 0$$

$$K2 = -\frac{1}{\sigma^2}\left[\left(\mu - \frac{1}{2}\sigma^2\right) + \sqrt{\left(\mu - \frac{1}{2}\sigma^2\right)^2 + 2r\sigma^2}\,\right] < 0$$

and h_1, h_2 are two constants to be determined.

The Firm's Market Value for Given S

By adding (13) and (16), equation (9), denoting the firm's market value for a given barrier S, can be rewritten as:

$$F(\theta_t; S, h_1, h_2) = M\theta_t^{\nu} + h_1(S)\theta_t^{K1} + h_2(S)\theta_t^{K2}, S \leq \theta_t < \infty \tag{18}$$

By definition, the first term on the right hand side represents the firm's market value when no control is exercised, whilst the term in brackets, i.e. the complementary function, may be interpreted as the value of the option to introduce a control (see Dixit 1991). This interpretation gives us some advice about the end-point conditions. If θ_t is very high, there is little probability that the lower barrier will be reached in a finite time-span. Therefore, the option of activating a barrier should be nearly worthless. For this, we need to set $h_1 = 0$. We can then omit the subscript on the remaining coefficient, and rewrite (18) as follows:

$$F(\theta_t; S, h) = M\theta_t^{\nu} + h(S)\theta_t^{K}, S \leq \theta_t < \infty \tag{18'}$$

The constant h remains to be determined, and this can be done by introducing a boundary condition on the general solution (18'). Since F is defined as the expected value of an integral of bounded cash flows, its pattern cannot be discountinuous except when θ_t hits S, and instantaneous control is exercised. In this case, where the regulator dL_t is of a small magnitude, and the cost dC_t is positive but infinitely small, the following boundary condition must hold:

$$F_\theta(S) = b \tag{19}$$

In other words, whatever the value of S, the firm pays bdS to exercise the control, and gets an increment of its market value equal to $F_\theta(S)dS$. The "smoothing pasting condition" (19) states that if $F(\theta_t; S)$ were not continuous and smooth at the critical point $\theta_t = S$, then the firm could do better by using the difference formed by the two terms in (19) to improve its market value.[5]

Written in terms of the functional (18'), condition (19) becomes $M\nu S^{\nu-1} + hKS^{k-1} = b$, giving

$$h = \frac{1}{K}\left[bS^{1-K} - M\nu S^{\nu-K}\right] \tag{20}$$

which can be substituted in (18'),

$$F(\theta_t; S) = M\theta_t^\nu + \frac{1}{K}\left[bS^{1-K} - M\nu S^{\nu-K}\right]\theta_t^K, S \leq \theta_t < \infty \tag{21}$$

Equation (21) represents the market value of a firm which carries out optimal current maintenance expenditure m_t, and exercises a control, at a marginal cost b, whenever θ_t hits S. Obviously, if no control is exercised, the firm's market value is given only by the particular solution (13).

3.3. *Conditions for Optimal Control*

Since S appears only in the constant $h(S)$, maximizing (21) with respect to S is equivalent to maximizing $h(S)$. In other words, optimizing $F(\theta_t; S)$ with respect to S is equivalent to finding a stationary value for $h(S)$.

After substitution of the general solution (21), condition (19) can be differentiated, keeping $h(S)$ constant. This gives the following condition:

$$F_{\theta\theta}(S^*) = 0 \tag{22}$$

In the literature dealing with instantaneous regulation of Brownian motion, equation (22) is referred to as "super contact condition" (Dumas, 1991), and is none other than the natural extension of (19) interpreted as the first order condition to determine the firm's optimal barrier, S^*. Since fixing a lower boundary S^* should do better, the constant h should be positive, i.e. the firm finds it profitable to restrict the admissibility range of θ_t downward. In fact, it can be shown that the second derivative of (19), evaluated at S^*, is strictly negative,

$$F_{\theta\theta\theta}(S^*) = \nu(\nu-1)(\nu-K)MS^{*\nu-3} < 0$$

which guarantees the existence of a maximum.

Solving (22) for S^*, we get:

$$S^* = \left[\frac{b(K-1)}{M\nu(K-\nu)}\right]^{1/(\nu-1)} \tag{23}$$

According to (23), the optimal barrier depends on the marginal cost of regulation (b), the maximal expected discounted profit coefficient (M), and the elasticity of the growth rate of the firm's profits with respect to θ, ν.[6]

Let us then consider the comparative statics of S^* with respect to σ^2 and b. With regard to the influence of the stochastic term σ^2, since it appears only in M and K, with

$$\frac{\partial M}{\partial \sigma^2} < 0, \quad \frac{\partial K}{\partial \sigma^2} > 0 \quad \text{for} \quad \mu > 0,$$

we get:

$$\frac{\partial S^*}{\partial \sigma^2} = \frac{1}{\nu-1}S^*\left[\frac{1}{M}\frac{\partial M}{\partial \sigma^2} + \frac{(\nu-1)}{(k-1)(K-\nu)}\frac{\partial K}{\partial \sigma^2}\right] < 0$$

In other words, the higher the uncertainty on the future evolution of the site quality, the more convenient it will be for the firm to push downward the lower boundary of the admissibility range for θ_t. Greater uncertainty in fact lowers the value of the option of the regulation, and consequently the firm will find it profitable to lower S^*. This result is attributable to the fact that the firm's instantaneous profit is a concave function of θ_t. Since with increasing uncertainty there is greater probability of θ_t hitting S, the firm will find it convenient to extend the range of variation in order to reduce the expected costs associated with site quality control.

Similar consequences derive from an increase in the marginal site quality regulation cost

$$\frac{\partial S^*}{\partial b} = -\frac{1}{(1-\nu)b}S^* < 0$$

i.e., higher values of b will induce the firm to choose lower levels of S.

4. THE AGENCY'S OPTIMIZATION PROBLEM

4.1. The Agency's Welfare Value for Any S and α

Substituting (14) in (5'), the agency's welfare function evaluated under the firm's optimal current maintenance expenditure rule becomes:

$$W(\theta_0; S, \alpha) = F(\theta; S) + \tag{24}$$

$$-E_0\left\{\int_0^{\infty} e^{-rt} D(\theta_t)dt + \rho\alpha b \int_0^{\infty} e^{-rt} dL_t \mid \theta(0) = \theta_0\right\}, S \leq \theta_t < \infty$$

where $F(.)$ is given by (21).

In order to obtain a closed form for (24), we then need to evaluate the remaining two integrals.

Let us start with the first one, which represents the total welfare loss attributable to the environmental impact of pollutant flows. To keep the problem mathematically tractable, we shall specify the damage function (2).

Assumption 6. $D(\theta_t) = p\theta_t^{\eta}, \qquad \eta \gtrless 0$

Given assumption 6, we can define the following functional:

$$TD(\theta_0; S) = E_0\left\{\int_0^{\infty} e^{-rt} p\theta_t^{\eta} dt \mid \theta(0) = \theta_0\right\}, S \leq \theta_t < \infty \tag{25}$$

which gives the expected discounted value of instantaneous damages over the entire planning horizon for given S.

Similarly, we can define the functional

$$TC(\theta_0; S) = E_0\left\{\int_0^{\infty} e^{-rt} b dL_t \mid \theta(0) = \theta_0\right\}, \quad S \leq \theta_t < \infty \tag{26}$$

which represents the total expected cost of controlling θ_t.

Proposition 3 *If the process $\{\theta_t\}$ is a regulated geometric Brownian motion as in (15), with starting value θ_0 at time zero, and a lower barrier at S, then:*

$$TD(\theta_t; S) = N\theta_t^{\eta} - \frac{1}{K}\left\{N\eta S^{\eta - K}\right\}\theta_t^K \tag{27}$$

where

$$N = \frac{p}{r - \mu\nu - \frac{1}{2}\nu(\nu - 1)\sigma^2} \equiv \frac{p}{\sqrt{B^2 - 4Aq}}$$

$$TC(\theta_t; S) = -b\left(\frac{1}{K}S^{1-K}\right)\theta_t^K \tag{28}$$

Proof: see Appendix B.

Notice that the first term on the right-hand-side of (27) represents the discounted value of environmental damage which is expected to result if no control on $\theta_t$is exercised, whilst the second term reflects the consequences, in terms of expected damage, of exercising such a control.

Substituting (27) and (28) in (24), the agency's welfare function becomes:

$$W(\theta_0; S, \alpha) = F(\theta_0; S) - TD(\theta_0; S) - \rho\alpha TC(\theta_0; S) \tag{29}$$

$$= M\theta_0^\nu - N\theta_0^\eta + \frac{1}{K}\left[(1+\rho\alpha)bS^{1-K} - M\nu S^{\nu-K} + N\eta S^{\eta-K}\right]\theta_0^K$$

The first two terms in (29) represent the agency's welfare value when no control is exercised. The remaining part may be interpreted as the value of the option of introducing a barrier S, such that the marginal welfare gain at $\theta_t = S$ is equal to the marginal social cost of pushing θ_t upward by an infinitesimal amount, i.e. $W_\theta(S) = (1 + \rho\alpha)b$.

4.2. *The Contract*

We may now proceed to the solution of the agency's control program (6), which we reproduce here for convenience:

$$\max_{s,\alpha} W(\theta_0; S, \alpha)$$

$$\text{s.t. } \{m_t\} = \arg\max\{F(\theta_0; S, \alpha), \text{ for given } S \text{ and } \alpha\},$$

$$F(\theta_0; S^*) \leq F(\theta_0; S, \alpha)$$

where $\{m_t\}$ is given by (14), $F(\theta_0; S^*)$ is given by (21) evaluated at S^* according to (23), and

$$F(\theta_0; S, \alpha) = M\theta_0^\nu + \frac{1}{K}\left[(1-\alpha)bS^{1-K} - M\nu S^{\nu-K}\right]\theta_0^K$$

represents the firm's market value associated with a reduction in the site quality regulation cost due to the subsidy offered in exchange for accepting the barrier indicated by the agency.

The first order conditions for an optimum are:

$$\frac{\partial W}{\partial S} = -\lambda \frac{\partial F}{\partial S} \tag{30a}$$

$$\frac{\partial W}{\partial \alpha} = -\lambda \frac{\partial F}{\partial \alpha} \tag{30b}$$

where λ is the Lagrangean multiplier associated with the participation constraint, and

$$\lambda[F(\theta_0; S, \alpha) - F(\theta_0; S^*)] = 0, \qquad \lambda \geq 0 \tag{30c}$$

From (30a) and (30b) we obtain the usual optimal condition

$$\frac{\frac{\partial W}{\partial S}}{\frac{\partial W}{\partial \alpha}} = \frac{\frac{\partial F}{\partial S}}{\frac{\partial F}{\partial \alpha}} \tag{31}$$

which states that, at the optimum, the agency's marginal rate of substitution between S and α should be equal to that of the firm. Recalling the expressions for $W(.)$ and $F(.)$, and rearranging (31), it can be easily shown that:

$$\frac{\frac{\partial W}{\partial S}}{\frac{\partial F}{\partial S}} = \frac{\frac{\partial W}{\partial \alpha}}{\frac{\partial F}{\partial \alpha}} \equiv -\rho \tag{32}$$

from which we get $\lambda = \rho > 0$, where ρ is the net social opportunity cost of public funds.

From (32), after some tedious manipulation, one may derive the equation which will give the agency's optimal barrier:

$$(1+\rho)(1-K)b - (1+\rho)(\nu - K)M\nu S^{\nu-1} + (\eta - K)N\eta S^{\eta-1} = 0 \tag{33}$$

According to (33), the optimality condition for S is independent of α, therefore the sufficient condition for a local maximum reduces to:

$$(1+\rho)(1-K)(\nu-1)b + (\eta - K)(\nu - \eta)N\eta S^{\eta-1} < 0 \tag{34}$$

which is always satisfied if:

$$\eta > 1, \text{ or}$$

$$0 < \eta < 1 \text{ and } (\eta - \nu) > 0, \text{ or}$$

$$\eta < 0 \text{ and } (\eta - k) > 0$$

The first case arises when instantaneous environmental damage is an increasing, convex function of θ_t. The second case occurs when environmental damage is again increasing but is a concave function of θ_t, while the growth rate of instantaneous profit is lower than that of instantaneous damage. Finally, in the third case, instantaneous damage is a decreasing function of θ_t, and its rate of decrease is lower than the decrease in the firm's market value attributable to regulating θ.

Equations (33) and (34) are non-linear in S. They yield some interesting analytical results, although numerical simulations are needed to get a better idea of the properties of the optimal solution.

The most important general result is that the agency's optimal barrier may be either higher or lower than the one the firm would choose in the absence of public intervention.

Proposition 4

If $\eta > 1$ *then* $S^{**} \leq S^*$;

if $0 < \eta < 1$ *and* $(\eta - \nu) > 0$ *then* $S^{**} \leq S^*$;

if $\eta < 0$ *and* $(\eta - K) > 0$ *then* $S^{**} \geq S^*$

where S^{**} *represents the agency's optimal barrier given by equation (33).*

Proof: see Appendix C.

The above results are not surprising. If environmental damage is positively correlated with the quality index θ_t, then the agency will find it convenient to offer the firm a subsidy that induces the latter to accept a wider admissibility range for θ. In other words, the firm will be induced to forego its own barrier and intervene only when θ_t hits a value lower than the one the firm would have chosen if bearing the full upkeeping costs of the site quality. By so doing, the agency will ensure a lower flow of expected environmental damages.

On the other hand, if $\eta < 0$, i.e. environmental damages are negatively correlated with site quality improvements, the agency will find it convenient to subsidize the firm so that it foregoes its own barrier and accepts a higher one.

Given S^*, obtained from solving equation (33), the optimal level of subsidization is obtained from the participation constraint,

$$\alpha = 1 - \Phi(S^{**}) \tag{35}$$

where:

$$\Phi(S^{**}) \equiv \frac{(1-\nu)}{(K-\nu)}\left(\frac{S^*}{S^{**}}\right)^{1-K} + \frac{M\nu}{b}(S^{**})^{\nu-1}$$

According to (35), the optimal rate of subsidization depends on the relative value of S^{**} with respect to S^*. Obviously if $S^{**} = S^*$, since $\Phi(S^{**}) = 1$, it would not be optimal for the agency to subsidize the firm's cost of controlling θ.

Proposition 5

If $S^{**} \geq S^*$ *then* $0 \leq \alpha \leq 1$;

if $S^{**} \leq S^*$ *then* $\alpha \geq 0$

Proof: see Appendix D.

This result is attributable to the asymmetry between the consequences of an increase and a decrease in S with respect to the firm's expected profits.

Let us first take the case in which the agency wishes to induce the firm to accept a higher barrier ($S^{**} \geq S^*$). As the agency is prepared to grant a subsidy which will lower the site quality regulation costs sustained by the firm, the latter will find it convenient to raise the level of the barrier with respect to the level it would have chosen if it had had to bear the entire lump sum maintenance cost. Because of the concavity of the profit function, the firm will be willing to accept higher values of S only if greater subsidies are provided: in other words, only if α increases. However, as the firm's market value increases with increases in S, it will not be necessary to grant a subsidy higher than the maintenance cost (i.e. $\alpha < 1$).

Let us now consider the case where $S^{**} \leq S^*$. Here, a subsidy is aimed at inducing the firm to accept a lowering of the barrier. This will bring about a reduction in the expected flow of the firm's marketable output which should be compensated by a reduction in the site quality upkeeping cost paid by the firm. However, for very low values of S, it will not be enough to grant the firm a subsidy which only partially covers regulation costs in order to compensate for its losses, but it will be necessary to provide a subsidy higher than the cost itself (i.e. $\alpha > 1$).

4.3. *Numerical Solutions*

The results proposed in the above Section may be better appreciated by means of numerical simulations. A separate examination of the two two polar cases described in described in propositions 4 and 5 is as follows.

Since we are primarily interested in exploring how the contract (S, α) offered by the agency is affected by uncertainty (σ) and the social opportunity cost of public funds (ρ), we will establish an arbitrary set of values for the remaining parameters: all the numerical simulations keep the discount rate constant $(r = 0.08)$; the production function parameters $(\nu = 0.4, q = 0.1)$; the site quality depreciation rate $(\delta = 0.10)$; the parameter p entering the environmental damage function $(p = 0.1)$; and the marginal cost of site quality upkeep $(b = 0.5)$.

Case 1: $\eta > 1$

Let us begin with the case where instantaneous environmental damage is an increasing, convex function of θ_t.

Setting $\eta = 3$, Table 1 shows how changes in σ and ρ would affect the contract offered by the agency (S^{**}, α).

Table 1.

	$\sigma = 0.30$	0.35	0.40	0.50
$\rho = 0.2$				
α	17.39	2.65	0.84	0.16
S^{**}	0.22	0.20	0.19	0.15
$\rho = 0.4$				
α	14.28	2.25	0.72	0.14
S^{**}	0.23	0.21	0.20	0.16
$\rho = 0.8$				
α	10.35	1.72	0.55	0.10
S^{**}	0.24	0.23	0.21	0.17
$\rho = 1.3$				
α	7.57	1.32	0.42	0.07
S^{**}	0.26	0.25	0.22	0.18
S^{*}	0.72	0.54	0.41	0.25

Case 2: $\eta < 0$ and $(\eta - K) > 0$

Let us now consider the case where the instantaneous damage is a decreasing function of θ_t.

Setting $\eta = -0.5$ we obtain the values shown in Table 2.

Table 2.

	$\sigma = 0.30$	0.35	0.40	0.50
$\rho = 0.2$				
α	0.39	0.32	0.25	0.09
S^{**}	2.15	1.48	1.02	0.43
$\rho = 0.4$				
α	0.36	0.29	0.23	0.08
S^{**}	2.01	1.38	0.95	0.41
$\rho = 0.8$				
α	0.32	0.25	0.19	0.06
S^{**}	1.80	1.24	0.86	0.38
$\rho = 1.3$				
α	0.28	0.21	0.16	0.05
S^{**}	1.62	1.12	0.79	0.35
S^{*}	0.72	0.54	0.41	0.25
K	−2.82	−1.73	−1.15	−0.62

5. CONCLUDING REMARKS

In this paper we have concentrated on the role of the physical features of the site where production activities take place in determining the extent of NPS discharges. We assumed that the state variable "site quality" is subject to a Brownian motion which may be regulated through a lump sum, linear upkeeping expenditure, and controls come into effect only when the site quality hits a predetermined lower barrier.

We first considered what would be the firm's optimal lower barrier without public intervention – i.e. when the firm bears the full site quality maintenance cost – and then concentrated on the environmental authority's problem, assuming that the agency is unable to continuously monitor the realization of the process driving the state variable. We then considered and analysed a policy under which firms will be granted a subsidy, proportional to lump sum expenditure for the upkeep of the site quality, in order to induce them to revise their decision on the site quality admissability range. This policy requires firms to inform the agency about the status of their site only when this reaches the minimum quality level indicated by the agency so as to receive the subsidy. In this way, site quality monitoring will be carried out only to ascertain that the claim is justified and the firm has taken the necessary steps required to improve quality.

Our main findings can be summarised as follows.

If the relationship between environmental damage and the site quality (θ_t) of the suspected polluter is such that the policy-maker wants the firm to raise the lower boundary of the admissability range of θ_t, S, the agency does not have to provide a subsidy higher than the cost of site quality regulation, and it would be inefficient to do so unless the social opportunity cost of public funds is equal to zero.

On the other hand, if the relation between environmental damage and the site quality index is such that the agency wishes to induce the firm to lower its minimum site quality threshold, it may happen that, for particularly low values of S, the offer of a subsidy lower than the lump sum maintenance cost will prove to be ineffective. That is, the firm will refuse the contract and will not review its choice of the admissability range of θ_t.

APPENDIX A

A demonstration of proposition 1 may be found in Karlin and Taylor (1981, ch. 15), and Harrison (1985, ch. 5); heuristic proof is given by Dixit (1991). In this appendix we will borrow the methodology used by Harrison for the case of a regulated Brownian motion as it is relevant for the development of this paper.

We assume that $\{\theta_t\}$ is a regulated homogeneous diffusion process, with initial condition $S \leq \theta_o < \infty$, satisfying the following condition:

i) $\{\theta_t\}$ is regulated with stochastic differential $d\theta_t = \mu(\theta_t)dt + \sigma(\theta_t)dz_t + dL_t$. The infinitesimal parameters $\mu(\theta_t)$ and $\sigma(\theta_t)$ are continuos functions of θ_t for $S \leq \theta_t < \infty$;
ii) $\{L_t\}$ is a finite variation, increasing and continous process;
iii) $\{L_t\}$ increases only when $\theta_t = S$;
iv) $\theta_t \in [S, \infty)$ for al $t \geq 0$.

The process $\{L_t\}$ maintains $\{\theta_t\}$ above the barrier S using the minimum amount of control; control takes place only when $\{\theta_t\}$ hits S. We then define the functional:

$$F(\theta_0) = E_0 \left\{ \int_0^\infty e^{-rt} \left[g(\theta_t)dt - bdL_t \right] \mid \theta(0) = \theta_0 \right\} \tag{A1}$$

By time-homogeneity of (A1), assuming the differentiability of F and applying the Itô's Lemma we obtain:

$$dF(\theta_t = F_\theta(\theta_t)d\theta_t + \frac{1}{2}F_{\theta\theta}(\theta_t)(d\theta_t)^2$$

$$= \left[F_\theta(\theta_t)\mu(\theta_t + \frac{1}{2}F_{\theta\theta}(\theta_t)\sigma^2(\theta_t) \right] dt + F_\theta(\theta_t)\sigma(\theta_t)dz_t + F_\theta(\theta_t)dL_t$$

The above equation accounts for the fact that, since the $\{L_t\}$ has finite variation, $(dL_t)^2 = 0$. Moreover, by the property (iii) $dL_t = 0$ except when θ_t= S. We thus obtain:

$$dF(\theta_t) = \left[F_\theta(\theta_t)\mu(\theta_t) + \frac{1}{2}F_{\theta\theta}(\theta_t)\sigma^2(\theta_t) \right] dt + F_\theta(\theta_t)\sigma(\theta_t)dz_t + F_\theta(S)dL_t \tag{A2}$$

Making use of the Integration By Parts proposed by Harrison (1985, p. 73) with $y_t = F(\theta_t)$ and $x_t = e^{-rt}$, and the fact that $dx_t = -rx_t$ and $dy_t$is given by (A2), we obtain:

$$e^{-rt}F(\theta_t) = F(\theta_0) + \int_0^t \left[\mu(\theta_\tau)F_\theta(\theta_\tau) + \frac{1}{2}\sigma^2(\theta_\tau)F_{\theta\theta}(\theta_\tau) - rF(\theta_\tau) \right] e^{-r\tau} d\tau$$

$$+\int_0^t \sigma(\theta_\tau)F_\theta(\theta_\tau)e^{-r\tau}dz_\tau + F_\theta(S)\int_0^t e^{-r\tau}dL_\tau$$

Now, provided that:

(a) $\lim_{u\to\infty} \Pr\left[T_u > T_a \mid \theta(0) = \theta_0\right] = 0$ for all $S \leq \theta_\tau < u < \infty$, where $T_a = \inf\{T \mid \theta_\tau = a\}$ is the minimum hitting time to $a \in [S, \infty)$;
(b) $\lim_{\theta\to\infty} F(\theta_\tau) = V(\theta_\tau)$;
(c) $\sigma(\theta_\tau)F_\theta(\theta_\tau)$ is bounded;
(d) $F_\theta(S) = b$;
(e) $\mu(\theta_\tau)F_\theta(\theta_\tau) + \frac{1}{2}\sigma^2(\theta_\tau)F_{\theta\theta}(\theta_\tau) - rF(\theta_\tau) = -g(\theta_\tau)$

taking the expectation at time zero of (A3) and letting t go to infinite we obtain (A1).

Condition (a) states that the probability of reaching infinite before a generic value a in $[S, \infty)$ is zero. Therefore $u \to \infty$ is a *non attracting* boundary for the process $\{\theta_\tau\}$ or, should it be attracting, it is an *unattainable* boundary, i.e. the mean time to reach it is infinite (Karlin-Taylor 1981, pp. 228–229).

Condition (b) states that as θ becomes very large, the probability that over any finite time period it should hit S becomes very low, and an active firm must have the expected value of $V(\theta)$ from the feasible strategy of never introducing the control (Karlin-Taylor 1981, pp. 228–229).

Finally from condition (e) and taking:

$$\mu(\theta_\tau) = \left[m(\theta_\tau)^\xi \theta_\tau^{-\gamma} - \delta\right]\theta_\tau,$$

$$\sigma(\theta_\tau) = \sigma\theta_\tau,$$

$$g(\theta_\tau) = q\theta_\tau^\nu - m(\theta_\tau).$$

we obtain (8) in the text, completing the proof of the proposition.

APPENDIX B

Keeping assumptions (i), (ii), (iii) and (iv) in Appendix A, expanding dTD using the Itô's Lemma, integrating By Parts as carried out in Appendix A, and rearranging, we obtain the following differential equation for $TD(\theta_t)$:

$$\frac{1}{2}\sigma^2\theta_t^2 TD_{\theta\theta}(\theta_t) + \mu\theta_t TD_\theta(\theta_t) - rTD(\theta_t) + p\theta_t^\nu = 0 \qquad \text{(B1)}$$

This equation must be solved subject to the same conditions as in Appendix A except (d) which is replaced by:

$$\text{(d')} \qquad TD_\theta(S) = 0$$

Again the procedure is to find a particular solution of $(B1)$ to be added to the solution of the homogeneous equation. In this case we can guess a functional form and determine by substitution whether it works. It is easy to see that the particular solution is of the form:

$$VTD(\theta_t) = N\theta_t^\nu \qquad \text{(B2)}$$

where N is a constant to be determined. Taking the partial derivative of $(B2)$ with respect to θ_t yields:

$$VTR_\theta = \nu\theta_t^{-1}VTR$$

$$VTR_{\theta\theta} = \nu(\nu-1)\theta_t^{-2}VTR$$

Substituting these last expressions into $(B1)$ and solving for the coefficient N, we get:

$$N = \frac{p}{r - \mu\nu - \frac{1}{2}\nu(\nu-1)\sigma^2} \equiv \frac{p}{\sqrt{B^2 - 4Aq}}$$

Omitting the last term on the r.h.s. of $(B1)$ it can be verified that the homogeneous equation is equal to the one obtained for the firm's market value. Thus, we can write the general solution of $(B1)$ as:

$$TD(\theta_t; S, h_1, h_2) = N\theta_t^\nu + h_1(S)\theta_t^{K1} + h_2(S)\theta_t^{K2} \qquad \text{(B3)}$$

where h_1 and h_2 are constants to be determined, and $K1, K2$ are the positive and negative roots of the quadratic equation $\frac{1}{2}\sigma^2K^2 + (\mu - \frac{1}{2}\sigma^2)K - r = 0.$

In addition, imposing the conditions (a) and (b) in Appendix A, and (d'), the constants become:

$$h_1 = 0 \qquad \text{and} \qquad h_2 = -\frac{N\nu}{K} S^{\nu - K} > 0,$$

where $K = K2$. Finally, using these values in $(B3)$ the first part of the proposition is proved.

Similarly, expanding dTC using the Itô's Lemma, integrating By Parts and rearranging the expressions yields the following differential:

$$\frac{1}{2}\sigma^2\theta_t^2 TC_{\theta\theta}(\theta_t) + \mu\theta_t TC_\theta(\theta_t) - rTC(\theta_t) = 0 \tag{B4}$$

This must be solved under the same conditions as in Appendix A except (b) and (d) which need to be replaced by:

(b') $TC(\theta_t)$ is bounded

(d'') $TC_\theta(S) = -b$

Condition (d'') refers to the benefits in terms of cost savings associated with regulation which are obtained by increasing θ by an infinitesimal amount.

Unlike the firm's market value and the environmental damage, $(B4)$ is only a homogeneous differential equation having the following solution:

$$TC(\theta_t; S, h_1, h_2) = h_1(S)\theta_t^{K1} + h_2(S)\theta_t^{K2} \tag{B5}$$

where $K1$ and $K2$ are the same roots as in the text. Now imposing (b') and (d'') and omitting the subscript from $K2$, we get:

$$h_1 = 0 \qquad \text{and} \qquad h_2 = -\frac{b}{K} S^{1-K} > 0$$

Using these values in $(B5)$ the proof is complete.

APPENDIX C

Let us consider equation (33) as a functional, calculate the first derivative, and the value when $S = S^*$.

$$\Psi(S) = (1+\rho)(1-K)b - (1+\rho)(\nu - K)M\nu S^{\nu-1} + (\eta - K)N\eta S^{\eta-1}$$

$$\frac{d\Psi}{dS} = -(1+\rho)(\nu - K)(\nu - 1)M\nu S^{\nu-2} + (\eta - K)(\eta - 1)N\eta S^{\eta-2}$$

$$\Psi(S^*) = (\eta - K)N\eta(S^*)^{\eta-1}$$

Let us examine separately the three cases examined in the proposition.

1) $\eta > 1$.

First, since the derivative is always positive, the function Ψ increases with S; second, $\Psi(S^*) > 0$; finally as $S \to 0$, $\Psi(S) \to -\infty$, whilst as $S \to +\infty$, $\Psi(S) \to +\infty$. Consequently, if there is a S^{**}such that $\Psi(S^{**}) = 0$, we get $S^{**} \leq S^*$.

2) $0 < \eta < 1$, and $(\eta - \nu) > 0$.

In this case the derivative is positive for $S < \bar{S}$, and negative otherwise, where $\Psi'(\bar{S}) = 0$.$\Psi(S^*)$ is still positive and as $S \to 0$, $\Psi(S) \to -\infty$, whilst as $S \to +\infty$, $\Psi(S) \to (1+\rho)(1-K)b > 0$. Consequently, if there is an S^{**}such that $\Psi(S^{**}) = 0$, we get $S^{**} \leq S^*$.

3) $\eta < 0$, and $(\eta - K) > 0$.

The derivative is always positive, and the function Ψ increases with S; $\Psi(S^*) < 0$; as $S \to 0$, $\Psi(S) \to -\infty$, whilst as $S \to +\infty$, $\Psi(S) \to (1+\rho)(1-K)b > 0$. Consequently, if there is a S^{**} such that $\Psi(S^{**}) = 0$, we get $S^{**} \geq S^*$.

APPENDIX D

From the participation constraint we get:

$$\alpha = 1 - \Phi(S^{**}),$$

$$\text{where: } = \Phi(S^{**}) = \frac{(1-\nu)}{(K-\nu)}\left(\frac{S^*}{S^{**}}\right)^{1-K} + \frac{M\nu}{b}(S^{**})^{\nu-1}$$

Let us start by considering the case in which $S^{**} = S^*$. Substituting the value for S^* given by (23) one can immediately verify that $\Phi = 1$, therefore $\alpha = 0$. We then consider the first and second derivative of the finction Φ evaluated at $S^{**} = S^*$.

$$\frac{d\Phi}{dS^{**}} \mid S^{**} = S^* = \frac{1-\nu}{K-\nu}(K-1)(S^*)^{-1} + \frac{M\nu}{b}(\nu-1)(S^*)^{\nu-2}$$

$$= \frac{1}{S^*}\left[\frac{1-\nu}{K-\nu}(K-1) + \frac{M\nu}{b}(\nu-1)\frac{b(K-1)}{M\nu(K-\nu)}\right] = 0$$

$$\frac{d^2\Phi}{d(S^{**})^2} \mid S^{**} = S^* = \frac{1-\nu}{K-\nu}(K-1)(K-2)(S^*)^{-2} + \frac{M\nu}{b}(\nu-1)(\nu-2)(S^*)^{\nu-3}$$

$$= \frac{(K-1)(1-\nu)}{(S^*)^2} < 0$$

This guarantees the existence of a minimum rate of subsidization, $\alpha = 0$, when $S^{**} = S^*$. Moreover, since $\lim_{S^{**}\to+\infty} \Phi(S^{**}) = 0$, if $S^{**} > S^*$, we get $0 < \alpha < 1$. Finally, since $\lim_{S^{**}\to 0^+} \Phi(S^{**}) = -\infty$, if $S^{**} > S^*$, we get $\alpha > 0$.

ACKNOWLEDGEMENT

Support from the Foundation ENI Enrico Mattei is gratefully acknowledged. We are grateful to Andrea Beltratti, Carlo Carraro and Jean-Jacques Laffont for their comments on a first draft, and Sergio Pastorello for his assistance with the numerical solutions. Related material was presented at the conference on "Agricultural policy and environment" sponsored by the University of Minnesota, the University of Padova and the Regional Agency for Agricultural Development of Veneto, held at Motta di Livenza (Treviso, Italy), June 1993.

NOTES

1. This asymmetry may occur where the difficulty of monitoring emissions at source arises from the number or spatial dispersion of emitters, but it is less likely to occur when it is attributable to the mode of conveyance and the environmental fate of NPS pollutants.
2. Optimal regulation of Brownian motion is a topic which is finding several economic applications. Examples are the work of Bentolila and Bertola (1990) dealing with employment decisions with hiring and firing costs; Dixit (1989), on entry and exit decisions in foreign markets; Pindick (1988), dealing with irreversible investment decisions of a firm facing uncertain demand or costs.
3. To verify that V, given by (13), satisfies (8), note that after having substituted (14) into (11) the latter becomes equal to (8). That is:

$$\left[\frac{1}{2}\sigma^2\nu(\nu-1)V+\left(\frac{1}{2}\nu M-\delta\right)\nu V-rV+\left(q-\frac{1}{4}\nu^2M^2\right)\frac{V}{M}\right]=0$$

$$\left[\frac{1}{2}\sigma^2\nu(\nu-1)-\delta\nu-r+\frac{q}{M}\right]V=0$$

$$\left[\frac{1}{2}\nu^2M^2-BM+q\right]\frac{V}{M}=0$$

$$[2AM^2-BM+D]\frac{V}{M}=0$$

   Since M is one of the roots of the quadratic equation inside the square brackets the equation is satisfied.
4. With a stochastic differential equation such as (3) there may be a positive probability that θ becomes infinite during the firm's planning period $[0,\infty)$. If this is the case then the integral in (4) does not converge, that is, the firm's market value does not exist. However, following the arguments in Vorst (1987) and Moretto (1991) it can be shown that, under the optimal maintenance policy (14), infinite is neither an attracting nor an attainable boundary and, therefore, the value of the firm will be always positive and bounded (see conditions a and b Appendix A).
5. A heuristic derivation of this condition, based on stochastic dynamic programming, can be found in Dixit (1991), and, in the case of instanteneous control, it holds for any arbitrarily fixed, and not necessarily optimal, barrier.
6. Substituting S^* into (20) the value for the constant h can be obtained. That is:

$$h=\frac{b(1-\nu)}{K(K-\nu)}(S^*)^{1-K}>0$$

   It can be noted that in the case of $\nu>1$ it would never be optimal to regulate site's quality ($h<0$).

REFERENCES

Bentolila, S., Bertola, G. (1990). "Firing Costs and Labour Demand: How Bad is Eurosclerosis?", *Review of Economic Studies*, 57, pp. 381–402.

Dixit, A. (1989). "Entry and Exit Decisions under Uncertainty", *Journal of Political Economy*, 97, pp. 620–638.

Dixit, A. (1991). "A Simplified Treatment of the Theory of Optimal Regulation of Brownian Motion", *Journal of Economic Dynamic and Control*, 15, pp. 657–673.

Dosi, C., Moretto, M. (1992). "Interventi di politica ambientale in condizioni di informazione asimmetrica: il caso dell'inquinamento da sorgenti diffuse", in G. Muraro (Ed.) "*Economia dell'informazione e economia pubblica*", Il Mulino, Bologna, pp. 219–244.

Dosi, C., Moretto, M. (1993). "NPS Pollution, Information Asymmetry, and the Choice of Time Profile for Environmental Fees", in C.S. Russel and J.F. Shogren (Eds.), *Theory, Modeling and Experience in the Management of Nonpoint-Source Pollution*, Kluwer Academic Publishers, Norwell MA., pp. 99–121.

Dumas, B. (1991). "Super Contact and Related Optimality Conditions", *Journal of Economic Dynamics and Control*, 15, pp. 675–685.

Griffin, R., Bromley, D. (1982). "Agricultural Runoff as a Nonpoint Externality", *American Journal of Agricultural Economics*, 64, pp. 547–552.

Harrison, J.M. (1985). *Brownian Motion and Stochastic Flow Systems*, John Wiley and Sons, New York.

Karlin, S., Taylor, H.M. (1981). *A Second Course in Stochastic Processes*, Academic Press, New York.

Moretto, M. (1991). "Physical Assets, Maintenance Policy and Optimal Behavior of the Firm", Quaderno n. 6, Dipartimento di Scienze Economiche, Universita' di Padova.

Pindyck, R. (1988). "Irreversible Investment, Capacity Choice, and the Value of the Firm", *American Economic Review*, 78, pp. 969–985.

Segerson, K. (1988). "Uncertainty and Incentives for Nonpoint Pollution Control", *Journal of Environmental Economics and Management*, 15, pp. 87–98.

Shortle, J.S., Dunn, J.W. (1986), "The Relative Efficiency of Agricultural Source Water Pollution Control Policies", *American Journal of Agricultural Economics*, 68, pp. 668–677.

Vorst, A.C.F. (1987). "Optimal Housing Maintenance under Uncertainty", *Journal of Urban Economics*, 21, pp. 209–227.

Xepapadeas, A. (1992). "Environmental Policy Design and Dynamic Nonpoint-Source Pollution", *Journal of Environmental Economics and Management*, 23, pp. 22–39

6. Incentives For Nonpoint Pollution Control

JAMES S. SHORTLE AND DAVID G. ABLER

The Pennsylvania State University, Department of Agricultural Economics and Rural Sociology, University Park, U.S.A.

1. INTRODUCTION

One of the central issues in environmental economics since the 1960s has been the relative efficiency of alternative instruments for pollution control. The traditional theoretical context in which the choice among instruments is examined presumes that polluters can control their emissions deterministically, emissions can be monitored accurately at little or no cost, and the contributions of each polluter to ambient concentrations are known. These assumptions are not usually characteristic of nonpoint pollution problems.[1] The fundamental characteristic of nonpoint pollution is that pollutants follow diffuse pathways from their sources to receiving environmental media. In addition, discharges of pollutants or precursors often cannot be observed directly. The nonpoint problem is therefore characterized by considerable uncertainty about the contribution of individual polluters to ambient concentrations. In many instances it may not be possible to identify who is or is not a contributor from a given set of candidates.

Serious research on the economic theory of nonpoint pollution control began with Griffin and Bromley's (1982) paper on the choice between standards and incentives applied to inputs or indirect measures of emissions. A number of advances have been made since by Segerson (1988), Xepapadeas (1992) and others.[2] In this paper we present a set of incentive structures for the class of nonpoint pollution problems in which the emissions of a pollutant are stochastic and unobservable, but the environmental agency has a probabilistic model of the impacts of polluter decisions on ambient concentrations. This class would include agricultural as well as some other important categories. Two of the

C. Dosi & T. Tomasi (eds.), Nonpoint Source Pollution Regulation: Issues and Analysis, 137–149.

incentive structures are firm-specific nonlinear taxes on a potentially broad set inputs used in production and pollution abatement. Both of these incentive structures are adaptations of mechanisms developed by Dasgupta, Hammond and Maskin (1980) for point source emissions that can be costlessly metered. The third is an adaptation of a mixed-instrument scheme developed by Roberts and Spence (1976) to the nonpoint pollution problem.

The firm-specific incentives presented in this paper are not offered as administratively or politically appealing instruments. Indeed, as will be evident they are sufficiently complex and information-intensive to raise significant questions about their practicality. The mixed instrument scheme is much simpler and more realistic. It is comparable to instruments used to control nonpoint pollution in practice (e.g., taxes and licenses on the use of polluting inputs such as pesticides and fertilizers) but offers potential efficiency gains. However, as we will see, optimal implementation of the mixed scheme could still be quite complex and administratively costly.

2. ANALYTICAL FRAMEWORK AND CHARACTERIZATION OF THE OPTIMUM

For heuristic purposes, consider the problem of an environmental agency that wishes to reduce agricultural nonpoint water pollution in a specific watershed.[3] The flow of pollutants from farms to receiving waters is referred to as runoff. The agency is uncertain about the water quality damage resulting from runoff, the effects of input decisions on runoff, future environmental conditions influencing runoff (e.g., weather), and the effects of changes in input use on farm profits. The farmers are assumed to be price-takers and risk-neutral. The farmers in the watershed as a group exert no significant influence on prices. In addition, we rule out all types of collusion among the farms.

The agency is unable to directly monitor runoff from farms but can formulate expectations conditional upon observations of farm resource allocation and other relevant variables using a probabilistic model of runoff.[4] The general form of the agency's runoff model is

$$\text{Prob }[r_i \leq r_i^*] = \text{Prob }[g_i(x_i, \lambda_i) \leq r_i^*] \tag{1}$$

where $r_i (i = 1, ..., n)$ is runoff from farm i, x_i is a vector of farm input decisions on the ith farm, $g_i(.)$ is a function relating runoff to farm input use, and λ_i is a vector of unknown physical parameters representing public uncertainty about the level of runoff for any given values of x_i, due to imperfect knowledge of the physical and chemical processes involved. The model and the distribution of the $\lambda_i (i = 1, ..., n)$ are common knowledge. We assume that $g_i(.)$ is continuous and at least twice continuously differentiable in x_i for any value of λ_i. The agency faces ex-ante uncertainty about the effects of input decisions on runoff because of its imperfect knowledge of the runoff function. It also faces ex-post uncertainty about the effects of changes in farm resource allocation on runoff, providing that it does not monitor runoff from the farms.

For simplicity, the agency's objective is taken to be maximization of the expected net benefits of nonpoint pollution abatement in the watershed. The economic cost of water pollution is given by $D(\Sigma_i r_i, \epsilon)$ where ϵ is an unknown parameter representing public uncertainty about the damage resulting from aggregate runoff.[5] The assumption that runoffs are summed in order to find damage costs (i.e., that runoff from one farm is a perfect substitute for runoff from another) is unrealistic but simplifies the notation considerably. The mechanisms we present could be generalized to encompass more complex settings. The damage cost function is convex and at least twice continuously differentiable in the runoff levels for any value of ϵ.

The maximum profit obtainable by the ith farm for any choice of inputs and weather conditions is given by the function $\pi^i(x_i, \theta_i)$ where θ_i is a vector of specialized private knowledge of the farm operation. This function is concave and at least twice continuously differentiable in x_i for any value of θ_i. While individual farmers know their profit functions (i.e., the value of θ_i), other farmers and the agency do not. The true value of θ_i is denoted $\hat{\theta}_i$.

Given the agency's knowledge of the runoff function and the costs of runoff, as well as the farmers' private knowledge, ex-ante efficient production in the watershed maximizes

$$J_1 = \sum_i \pi^i(x_i, \hat{\theta}_i) - E\left[D\left(\sum_i g_i(x_i, \lambda_i), \epsilon\right)\right]. \tag{2}$$

The Kuhn-Tucker conditions are:

$$\frac{\partial J_1}{\partial x_{ij}} = \partial \pi^i / \partial x_{ij} - E\left[\partial D / \partial r(.)(\partial g_i / \partial x_{ij})\right] \leq 0, (\partial J_i / \partial x_{ij}) x_{ij} = 0, \tag{3}$$

and $x_{ij} \geq 0$, where x_{ij} is the jth element of the vector x_i.

Since the expected value of the product of two random variables is the product of their expected values plus the covariance, (3) can also be written

$$\partial\pi^i/\partial x_{ij} \leq E\,[\partial D/\partial r]\;E\,[\partial g_i/\partial x_{ij}] + COV(\partial D/\partial r(.), \partial g_i/\partial x_{ij}). \quad (4)$$

The left-hand side of (4) is the expected gain in profit at the margin from the jth input on the ith farm. In the absence of pollution controls, expected profit-maximizing farmers would only choose inputs levels with non-negative marginal effects on profits. The right-hand side of (4) is the expected marginal damage cost of the jth input on the ith farm. This is the sum of a term involving the impact of a change in the mean level of runoff and a term involving the impact of a change in the variability of runoff.

3. FIRST-BEST INCENTIVES

With perfect information about farm profits, the agency could obtain the optimal solution described above by imposing farm-specific design standards requiring each farm to adopt an optimal production plan. Alternatively, the agency could obtain the optimal solution by imposing a system of farm-specific linear input taxes of the form

$$L_i = \sum_j t_{ij} x_{ij} \quad (5)$$

where t_{ij} is the RHS of (4) evaluated at the optimal solution. Note that the RHS of (4) may be negative for some inputs. Hence, the tax rates could be negative for some inputs to encourage their use. Abatement inputs with negative marginal effects on profits would fall into this class, but other inputs that are normally used may also reduce runoff.

Implementing this approach requires that the agency possess the farmers' private knowledge. Hence, the problem is to obtain this knowledge.[6] If the dimension of the vector θ_i were less than or equal to the dimension of the vector x_i, it would be possible in principle for the public environmental agency to deduce θ_i from observed input choices (x_i), since it knows the form of the profit function. However, the administrative costs of this would probably be high. If the dimension of θ_i were greater than the dimension of x_i, it might still be possible to deduce θ_i using observed input choices for more than one time period (assuming that private information is not also changing over time). However, the administrative costs would again be high. Moreover, in either case, producers might be induced to deliberately distort input choices in order to mislead the government.

Alternatively, it is possible in principle to construct a set of firm-specific taxes that induce farmers to reveal their private knowledge and thus obtain the first-best solution. This structure is an adaptation of an incentive-compatible emissions tax scheme developed by Dasgupta et al. (1980) for non-stochastic emissions that are readily observable by source. Under the extension developed here, the agency asks farmers to provide their specialized knowledge. The farmers are informed that their reports will be used to formulate farm-specific tax schedules according to guidelines announced to the farmers. The guidelines are such that (a) the first-best solution is attained when all farmers maximize their after-tax profits, provided that each has given a true report, and (b) truthfulness serves the farmers' economic self-interest. The main distinction between this tax scheme and that of Dasgupta et al. (1980) is that the taxes are based on the farmers' input decisions rather than observed emissions.

Let θ_i' be the specialized information reported by the ith farmer. For any given values of the random variables λ_i and ϵ, let the set of vectors $x_i'(i = 1, ..., n)$ provide the maximum of

$$J_2 = \sum_i \pi^i\left(x_i, \theta_i'\right) - E\left[D\left(\sum_i g_i(x_i, \lambda_i), \epsilon\right)\right]. \tag{6}$$

Hence, the vectors x_i' will be the first-best solution if all farmers make true reports (i.e., $\theta_i' = \hat{\theta}_i$). When reporting its specialized information, the ith farm is informed that its tax will be

$$T_i' = E\left[D\left(g_i(x_i, \lambda_i) + \sum_{j \neq i} g_j(x_j', \lambda_i), \epsilon\right)\right] - \sum_{j \neq i} \pi_j(x_j', \theta_j') - c_i', \tag{7}$$

where c_i' is a lump sum subsidy (or, if negative, tax) chosen to prevent socially inefficient exit from or entry into the industry.

The ith farmer's after-tax profit function with the imposition of such a tax is $\pi^i(x_i, \hat{\theta}_i) - T_i'$. The construction of the tax is such that x_i' maximizes expected after-tax profits if and only if the farmer has been truthful. Knowing this, and not being able to influence the tax schedule it will ultimately face, each farm's dominant strategy is to be truthful and to choose x_i'. This mechanism is a dominant strategy mechanism.

The tax facing each farm depends not only on its input choices, but also on the socially optimal input levels on other farms (which depend on their private knowledge). In practice, an iterative scheme would probably have to be used, with initial tax schedules based on initial estimates of socially optimal input levels. The figures on specialized information given to the environmental agency (θ_i') would then be used to compute revised tax schedules, which would lead to revised reports of specialized information, and so on.

4. A SECOND BEST INCENTIVE

If the number of farmers is large, then the scheme above will require the transmission and processing of a lot of information in both the formulation of the farm-specific tax function and subsequent monitoring of farm management. The cost of this would almost certainly make the scheme uneconomic. On the other hand, if the number of farmers is small, the scheme may not achieve the first-best solution because of collusion: Farmers might collectively misrepresent their private information, making the negative impacts of input taxes on farm profits appear too large.

In this section, an alternative incentive structure is presented. This structure is a set of farm-specific taxes that maximizes the expected net benefit of production subject to the constraint that the farmers' specialized knowledge remains private. However, the government can still observe input choices. Of course, such a scheme cannot attain the first-best solution, but it does reduce the administrative cost of pollution control. Like the first-best scheme, this tax structure is an extension of an emissions tax scheme presented in Dasgupta et al. (1980) for non-stochastic emissions that are monitorable by source.

Suppose that, for the environmental agency, the θ_i are independently distributed random variables with known density functions. Suppose that this is also true for the farmers, except of course that they know their own $\hat{\theta}_i$. The λ_i are independently distributed with known density functions. Under this scheme, the tax function for the ith farm is

$$\bar{T}_i(x_i) = E\left[D\left(g_i(x_i, \lambda_i) + \sum_{j\neq i} g_j(\bar{x}_j(\theta_j), \lambda_j), \epsilon\right)\right]$$

$$- E\left[\sum_{j\neq i} \pi_j(\bar{x}_j(\theta_j), \theta_j)\right] - \bar{c}_i. \tag{8}$$

where $\bar{x}_j(\theta_j)$ is the solution to

$$\max_{x_j} \pi^j(x_j, \theta_j) + E\left[\sum_{i \neq j} \pi^i(\bar{x}_i(\theta_i), \theta_i) - D\left(g_j(x_j, \lambda_j) + \sum_{i \neq j} g_i(\bar{x}_i(\theta_i), \lambda_i)\right)\right] \tag{9}$$

Once again, $\bar{c}_i$ is a lump sum subsidy or tax chosen to prevent socially inefficient exit from or entry into the industry. The expectations in (9) use the government's information set, which now excludes farmers' private information. Like the first-best scheme, the tax facing each farm here depends not only on its input choices, but also on input choices by other farms. An iterative scheme to set the schedules would probably have to be used here as well.

Maximizing profits net of the tax in (8) is equivalent to maximizing

$$J_{3i} = \pi^i(x_i, \hat{\theta}_i) - E\left[D\left(g_i(x_i, \lambda_i) + \sum_{j \neq i} g_j(\bar{x}_j, \lambda_j), \epsilon\right)\right] + E\left[\sum_{j \neq i} \pi_j(\bar{x}_j, \theta_j)\right]. \tag{10}$$

In this sense, the tax scheme in (8) provides the maximum expected net social benefit from runoff control, subject to the constraint that the farmers' specialized knowledge remains private.

Although this scheme requires less information than the first-best scheme, administrative costs are still likely to be formidable. The agency must observe input choices for each farm, formulate expectations about each farm's private information, and set tax schedules for each producer that will likely need to be fine-tuned in an iterative manner.

5. A SYSTEM OF MIXED INSTRUMENTS

The first two incentive structures are firm-specific nonlinear taxes on a potentially broad base of inputs influencing the flow of pollutants from the firm. These incentives are sufficiently complex and information-intensive, even in the case of the second-best incentive, to raise significant questions about their practicality. Further concessions to administrative and compliance costs would involve uniform (at least within a given area) rather than firm-specific input tax structures, linear as opposed to nonlinear tax schedules, truncation of the input base to inputs that are easily monitored while also being key polluting or pollution-abatement inputs, and specification of policy parameters by means other than cost-benefit assessments. Examples are the kinds of instruments that are actually observed in practice. These include taxes on polluting inputs such as fertilizers and pesticides, limits (licenses) on the acceptable levels of use of such inputs, and design standards. Corresponding to the reduction in administrative and compliance costs would be a loss in efficiency. The mixed instrument scheme considered below is consistent with the input-based instruments that are used in practice but represents a potential improvement over conventional approaches.[7]

The system consists of taxes, subsidies and permits for the use of inputs (eg. fertilizers, pesticides, etc.) that influence polluting discharges. It is an adaptation of a scheme proposed by Roberts and Spence (1976). However, in our analysis the instruments are applied to inputs rather than emissions, which we continue to assume are unobservable. In addition, whereas Roberts and Spence assume that the public sector knows the cost functions of the firms up to a common parameter, we allow for specialized knowledge on the part of each firm. Farms must hold licenses to use polluting inputs. The licenses can be traded. However, unlike the usual license system, farms may use more inputs than allowed by the permits that they hold provided that they pay a tax on the additional amount. Conversely, a farm may receive a subsidy if it returns permits to the public sector.

To simplify our analysis we assume only two inputs and a single private information variable for each farm. To add some realism to our story, we assume that the scheme is applied only to input 1. This may be because input 1 is easily observed while 2 is not, or because 2 has no direct influence on the flow of discharges. If 2 has no direct influence, then there is no efficiency loss from not applying an instrument to this variable. If it does have a direct influence, then there is an efficiency loss. The market price of permits is p, the amount of permits held by farm i is z_i, the tax rate on excess use is t and the subsidy rate for returned permits is s. The tax rate exceeds the subsidy rate. The supply of permits is z. In competitive permit market equilibrium it will be the case that

$$\sum z_i = z \tag{11}$$

In such an equilibrium it must be the case that $s \leq p \leq t$. If $t \leq p$ then farms could expand input use more cheaply by paying the tax than buying permits to avoid the tax. The demand for permits would collapse and the price would fall at least to the tax rate. Accordingly, the tax sets the upper limit on the equilibrium permit price. Alternatively, if $p \leq s$ then permits could be "sold" back to the government for a higher price (i.e., the subsidy rate) than they could be purchased on the market. With arbitrage the permit price would be driven at least to the subsidy rate. Hence the subsidy rate sets the lower limit for the permit price. The equilibrium permit price will be a function of the specialized information of the farmers and the permit supply. We denote the equilibrium price as $p(\theta, z)$ where $\theta = [\theta_i]$.

The ith farmer maximizes

$$\pi^i(x_{i1}, x_{i2}, \theta_i) + s(z_i - x_{i1}) - pz_i \tag{12}$$

if the profit maximizing level of x_{i1} is less than z_i. In this case the optimal values of x_{i1} and x_{i2} solve the system

$$\frac{\partial \pi^i}{\partial x_{i1}} = s, \frac{\partial \pi^i}{\partial x_{i2}} = 0 \tag{13}$$

We will assume this to be the solution when θ_i is less than or equal to $\underline{\theta}_i$. The optimal values of the inputs for this case are $\underline{x}_{i1}(s, \theta_i)$ and $\underline{x}_{i2}(s, \theta_i)$. Alternatively, the farmer maximizes

$$\pi^i(x_{i1}, x_{i2}, \theta_i) - t(x_{i1} - z_i) - pz_i \tag{14}$$

if the profit maximizing level of x_{i1} is greater than or equal to z_i. In this case the optimal values of x_{i1} and x_{i2} solve the system

$$\frac{\partial \pi^i}{\partial x_{i1}} = t, \frac{\partial \pi^i}{\partial x_{i2}} = 0. \tag{15}$$

We will assume this to be the solution when θ_i is more than or equal to $\bar{\theta}_i$. The optimal values of the inputs for this case are $\bar{x}_{i1}(t, \theta_i)$ and $\bar{x}_{i2}(t, \theta_i)$. For $\underline{\theta}_i \leq \theta_i \leq \bar{\theta}_i$, the farmer will hold as many permits as required by the optimal value of x_{i1} but no more. The optimal values of x_{i1} and x_{i2} in this case solve the system

$$\frac{\partial \pi^i}{\partial x_{i1}} = p, \frac{\partial \pi^i}{\partial x_{i2}} = 0. \tag{16}$$

The optimal values of the inputs for this case are $\hat{x}_{i1}, (p, \theta_i)$ and $\hat{x}_{i2}(p, \theta_i)$. Given $s \leq p \leq t$, it will be the case that

$$\underline{x}_{i1}(s, \theta_i) \leq \hat{x}_{i1}(p, \theta_i) \leq \bar{x}_{i1}(t, \theta_i) \tag{17}$$

Unlike the structures considered above, this mixed system can be implemented without the use of information on farm profits or damage costs. Of course, no claims for the efficiency of such a structure can be made. To determine the optimal set of parameters of the structure, the problem is to choose a tax rate t, permit supply z and subsidy rate s to maximize

$$\sum_i^n \int_0^{\underline{\theta}_i} \left[\pi^i(\underline{x}_{i1}(s,\theta_i), \underline{x}_{i2}(s,\theta_i), \theta_i)\right] f_i(\theta_i)d\theta_i \tag{18}$$

$$-\prod_i^n \int_{\lambda_i} \int_0^{\underline{\theta}_i} \left[D\left(\sum_i^n g_i(\underline{x}_{i1}(s,\theta_i), \underline{x}_{i2}(s,\theta_i), \lambda_i)\right)\right] \prod_i^n f_i(\theta_i)k_i(\lambda_i)d\theta_i d\lambda_i$$

$$+\sum_i^n \prod_i^n \int_{\underline{\theta}_i}^{\bar{\theta}_i} \left[\pi^i(\hat{x}_{i1}(p(\theta,z),\theta_i) - \hat{x}_{i2}(p(\theta,z),\theta_i),\theta_i)\right] \prod_i^n f_i(\theta_i)k_i(\lambda_i)d\theta_i d\lambda_i$$

$$-\prod_i^n \int_{\lambda_i} \int_{\underline{\theta}_i}^{\bar{\theta}_i} \left[D\left(\sum_i^n g_i(\hat{x}_{i1}(p(\theta,z),\theta_i), \hat{x}_{i2}(p(\theta,z),\theta_i), \lambda_i)\right)\right]\Bigg) \prod_i^n f_i\,(\theta_i)\,k_i\,(\lambda_i)\,d$$

$$+\sum_i^n \int_{\bar{\theta}_i}^{b_i} \pi^i(\bar{x}_{i1}(t,\theta_i), \bar{x}_{i2}(t,\theta_i), \theta_i)\Big]\, f_i(\theta_i)d\theta_i$$

$$-\prod_i^n \int_{\lambda_i} \int_{\bar{\theta}_i}^{b_i} \left[D\left(\sum_i^n g_i(\bar{x}_{i2}(t_i,\theta_i), \bar{x}_{i2}(t,\theta_i), \lambda_i)\right)\right] \prod_i^n f_i(\theta_i)k_i(\lambda_i)d\theta_i d\lambda_i$$

where $f_i(\theta_i)$ is again the public decision-maker's pdf. for θ_i, $k_i(\lambda_i)$ is the public decision maker's pdf. for λ_i, and b_i is the upper limit of θ_i.

The first order conditions are:

$$\sum_i^n E\left[\left(s - \frac{\partial D}{\partial r}\frac{\partial g_i}{\partial x_{i1}}\right)\frac{\partial \underline{x}_{i1}}{\partial s} + \left(\frac{\partial \pi^i}{\partial x_{i2}} - \frac{\partial D}{\partial r}\frac{\partial g_i}{\partial x_{i2}}\right)\frac{\partial \underline{x}_{i2}}{\partial s}\right. \tag{19}$$

$$\left.\mid \theta_i \le \underline{\theta}_i \text{ for all } i = 1, ..., n\right] = 0$$

$$\sum_i^n E\left[\left(\left(p - \frac{\partial D}{\partial r}\frac{\partial g_i}{\partial x_{i1}}\right)\frac{\partial \hat{x}_{i1}}{\partial p} + \left(\frac{\partial \pi^i}{\partial x_{i2}} - \frac{\partial D}{\partial r}\frac{\partial g_i}{\partial x_{i2}}\right)\frac{\partial \hat{x}_{i2}}{\partial p}\right)\frac{\partial p}{\partial z}\right. \tag{20}$$

$$\left.\mid \underline{\theta}_i \le \theta_i \le \bar{\theta}_i \text{ for all } i = 1, ..., n\right] = 0$$

$$\sum_i^n E\left[\left(t - \frac{\partial D}{\partial r}\frac{\partial g_i}{\partial x_{i1}}\right)\frac{\partial \bar{x}_{i1}}{\partial t} + \left(\frac{\partial \pi^i}{\partial x_{i2}} - \frac{\partial D}{\partial r}\frac{\partial g_i}{\partial x_{i2}}\right)\frac{\partial \bar{x}_{i2}}{\partial t}\right. \tag{21}$$

$$\left.\mid \theta_i \geq \bar{\theta}_i \text{ for all } i = 1, ..., n \right] = 0$$

The mixed instruments obtained by solving these conditions must perform at least as well as a policy composed of the individual components when formulated with the same information. If a linear tax alone is optimal, then it would be a special case of the mixed instrument system with $s^* = -t^*$ and $z^* = 0$ (where *'s denote optimal values). Alternatively, if a permit system alone is optimal, then it would be a special case of the mixed instrument system with $s^* = t^* = 0$. In general, however, these limiting forms would not be the optimal specification. Hence, in general this mixed system will dominate policies formulated from the individual components.

There are several modifications that might improve the performance of the mixed instrument structure presented here. One is to specify a tax, subsidy and permits for the second input. The system we have specified involves adjustments in the values of the parameters of the system to account for the effects of the choice of the parameters for the monitored input on the unmonitored input. Defining a mixed system for the second input would be efficient if x_{i2} entered $g^i(.)$ and if gains from the additional structure were expected to exceed the increased costs of monitoring and enforcement. In addition, additional steps could be made in the marginal cost of using polluting inputs.

This mixed system economizes on information costs relative to the two schemes listed above, since the allocation of polluting inputs among firms is left to the private sector via a price system. Unlike the first two schemes, no one needs to calculate the cost of reducing polluting inputs on each farm. A farm merely needs to compare its cost, which it knows anyway, with the market price of permits. However, the government still needs to determine the total quantity of permits to be issued, as well as the tax rate on excess use and the subsidy rate for returned permits. Like the first two schemes, this requires estimation of runoff levels and damage costs.

6. CONCLUDING REMARKS

Three incentive structures have been presented for the general nonpoint pollution control problem in which emissions are stochastic and unobservable at reasonable cost at the firm and more aggregate levels, and there is differential information about the costs of control. One is incentive compatible and capable of achieving the ex-ante efficient outcome, while the other two economize on information with a corresponding loss of efficiency. The third is the most practical of the three. It involves marketable permits for inputs associated with nonpoint pollution, combined with a tax on excess use and a subsidy for returned permits. Optimal implementation could still entail large administrative costs, but the structure should offer opportunities for increased efficiency over the types of input-based tax and license schemes observed in practice.

NOTES

1. An excellent survey of the problems encountered in nonpoint pollution control from an economic perspective is found in Braden and Segerson (1991).
2. Braden and Sergerson (1991) present a thorough survey of the theoretical literature on nonpoint pollution control.
3. Nonpoint water pollution is largely associated with runoff from open areas such as agricultural land, forest land, mine sites, and paved areas. Agricultural land is a leading nonpoint source of surface and groundwater pollution in the U.S and many other countries (OECD, 1989). For simplicity, we assume a single pollutant.
4. Models for estimating agricultural nonpoint pollution flows from farm fields and watersheds based on observations of land use practices, topography, weather, and other relevant factors have been developed and continue to be refined (DeCoursey, 1985). Such models reduce but do not eliminate uncertainty about nonpoint pollution flows from individual farms.
5. In many cases too little may be known about the damage cost function to make use of it in policy design. The analysis presented here will hold for a second-best criterion that substitutes a penalty function obtained through some legislative or administrative process for the unknown damage cost function.
6. In general, it is reasonable to assume that firms have better information about the effects of changes in production on their profits than the public planner. Such a differential information structure is often assumed in the literature on choices among pollution policy instruments.
7. Another alternative which we do not address here is the application of taxes or standards to indirect measures of emissions. Some of the issues that are involved are addressed in Shortle and Dunn (1986) and Shortle (1987).

REFERENCES

Braden, J., Segerson, K. (1991). *Information Problems in the Design of Nonpoint Source Pollution Policy*, Presented at the AERE workshop on Management of Nonpoint Source Pollution, Lexington, KY, June.

Dasgupta, P., Hammond, P., Maskin, E. (1980). "On Imperfect Information and Optimal

Pollution Control", *Review of Economic Study*, 47, pp. 857–60.

DeCoursey, D. (1985), "Mathematical Models for Nonpoint Pollution Control", *Journal of Soil and Water Conservation*, 40, pp. 408–413.

Tomasi, T., Segerson, K., Braden, J. "Issues in the Design of Incentives Schemes from Nonpoint Source Pollution Control." (This volume).

Griffin, R., Bromley, D. (1982), "Agricultural Runoff as A Nonpoint Externality", *American Journal of Agricultural Economics*, 64, pp. 547–52.

Organization for Economic Cooperation and Development (1989). *Agricultural and Environmental Policies: Opportunities for Integration*, OECD, Paris.

Roberts, M., Spence, M. (1976). "Effluent Charges and Licenses under Uncertainty", *Journal of Public Economics*, 5, pp. 193–208.

Segerson, K. (1988). "Uncertainty and Incentives for Nonpoint Pollution Control", *Journal of Environmental Economics and Management*, 15, pp. 87–98.

Shortle, J., Dunn, J. (1986). "The Relative Efficiency of Agricultural Source Water Pollution Control Policies", *American Journal of Agricultural Economics*, 68, pp. 668–677.

Shortle, J. (1987). "The Allocative Implications of Comparisons Between the Marginal Costs of Point and Nonpoint Source Pollution Abatement", *Northeastern Journal of Agriculture and Research in Economics*, 17, pp. 17–23.

Xepapadeas, A. (1991). "Environmental Policy Under Imperfect Information: Incentives and Moral Hazard", *Journal of Environmental Economics and Managegement*, 20, pp. 113–126.

7. Water Pollution Regulation in a Spatial Model

THEODORE TOMASI* AND ARTHUR WEISE**

**Department of Agricultural Economics, Michigan State University;*
***Agricultural Research Service, U.S. Department of Agriculture*

1. INTRODUCTION

In this paper the properties of optimal tax incentives schemes to control pollution in the long run in a spatial setting are examined. In the full information, "point source" setting, a majority of authors who have investigated this issue attest to the efficacy of spatial Pigouvian taxes (e.g., Tietenberg, 1974; Mäler, 1974; Hochman *et al.*, 1977; Freeman, 1984; Spulber, 1985). Such taxes are levied on emissions, at a rate equal to the marginal damages of emissions at that location, evaluated at an optimal allocation. However, close examination of these papers reveals that, while excellent contributions in a number of respects, they fail to demonstrate completely that a system of spatial Pigouvian taxes (SPTs) suffice to ensure that competitive equilibrium (given the taxes) are Pareto efficient. The reason is that they fail to consider adequately the assignment of productive activities to locations.

Tietenberg required that each production zone have at least one emitter, while Freeman required all firms to be at an interior solution. Baumol and Oates follow Freeman's analysis. Clearly, models which assume interior solutions are not capable of addressing the issue of which locations should be inactive in an efficient allocation. Hochman *et al.* (1977) study a continuous space model similar to Henderson (1977) and that investigated below, but do not choose optimally the boundary of productive activities and residential land uses; rather they assume that land rents from these uses are equal in an efficient equilibrium.

C. Dosi & T. Tomasi (eds.), Nonpoint Source Pollution Regulation: Issues and Analysis, 151–174.

However, since their transition equation satisfies our condition (discussed below) for their conclusion to hold, their paper is not in error. Mäler (1974) and Spulber (1985) study non-spatial, long-run models and claim, without proof, that their results on the efficacy of emission taxes extend to a spatial setting.

In a model specifically devoted to the question of location in an equilibrium setting, Henderson (1977) offers a finding contrary to those described above. He writes (p. 90):

> "...in a spatial world, optimal environmental policies no longer consist of just imposing an appropriate set of Pigouvian taxes. ... neighborhood externalities or distortions in the urban land market are created when the spatial configuration of firms affects the pollution disamenities suffered by residents. These distortions require additional intervention in the market to effect an optimal allocation of land between residential and commercial uses."

Here, the issues raised by Henderson are studied, and an attempt is made to reconcile his findings with the prevailing wisdom that SPTs suffice to internalize spatial externalities. This paper provides a condition under which the "additional intervention" Henderson calls for is not needed and the conclusions of the previously cited papers are upheld. The sufficient condition deduced here restricts the transition equation governing accumulation of pollution through space. The restriction is that the transition equation is specified to be linear (not affine) in the rate of inflow of new pollutants from production activities and separable between this inflow and accumulated pollution.

A result similar to that of Henderson was derived by Carlton and Loury (1980, 1986) in a non-spatial setting. They note that in the long run, in addition to the usual tax on emissions, a tax may be required (depending on the form of the environmental damage function) to control entry decisions. It might be reasoned that Henderson's proposed intervention governing the location of production in a spatial setting is the same as Carlton-Loury tax on the number of firms. We show that, while the two are closely related, they are not necessarily the same.

We then turn briefly to issues of "nonpoint source" pollution, (reviewed by Braden and Segerson; and Tomasi *et al.*, Chapter 1, this book), where uncertainty and asymmetric information aspects of the pollution control problem are crucial. The analysis in this setting is much less well-developed. Here, we sketch some implications of the point source results discussed previously for the nonpoint, moral hazard incentive schemes studied in the literature to date, and suggest a method for investigating these concerns further in the basic model that we study here. Location issues in nonpoint models have been addressed by Laffont (Chapter 2, this book), and by Moretto and Graham-Tomasi (1992).

2. THE MODEL

Our model combines features of those formulated by Henderson (1977), by Hochman *et al.* (1977), and by Carlton and Loury (1980, 1986). It builds on earlier spatial equilibrium models of Muth (1961) and Solow and Vickery (1971), but incorporates pollution effects. The model concerns water pollution generated by agricultural activities, but easily could be adapted to air pollution, with which Henderson is concerned. Moreover, we are concerned more generally with the literature on the control of point sources of pollution. We ignore the important informational problems associated with nonpoint pollution from agricultural lands.

We study a rectangular region in (t, s)-space. In the t-direction runs a river, and on land along the river an agricultural crop is grown in an interval $[0, S(t)]$ in the s-direction. At each t, farmland availability is limited by a constraint $S(t) \leq S^o$. In the t-dimension, farming takes place between $t = 0$ and $t = t^f \leq T$ Agricultural production generates a pollutant which is deposited in the river and transported to a lake located at $t = T$. As the pollutant moves in the river, the pollution's impact on environmental amenities in the lake is ameliorated.

Consumers/laborers/landlords live in a residential area between $t = t^r$ and $t = T$. Consumers provide labor to production of a manufactured good at $t = T$, purchase the agricultural good and manufactured goods, and enjoy lake water quality. Consumers incur commuting costs to get to the central business/lake district locate at $t = T$. Their income is derived from wages, agricultural land rents, and pollution tax receipts.

The region is depicted in Figure 1. We now flesh out the details of the model.

Agricultural Production

Let $x(t, s)$ be the density of application of an agricultural input at location (t, s), where $x(t, s) > 0$ if $(t, s) \in [0, t^f]X[0, S(t)]$ and $x(.) = 0$ otherwise. Thus,

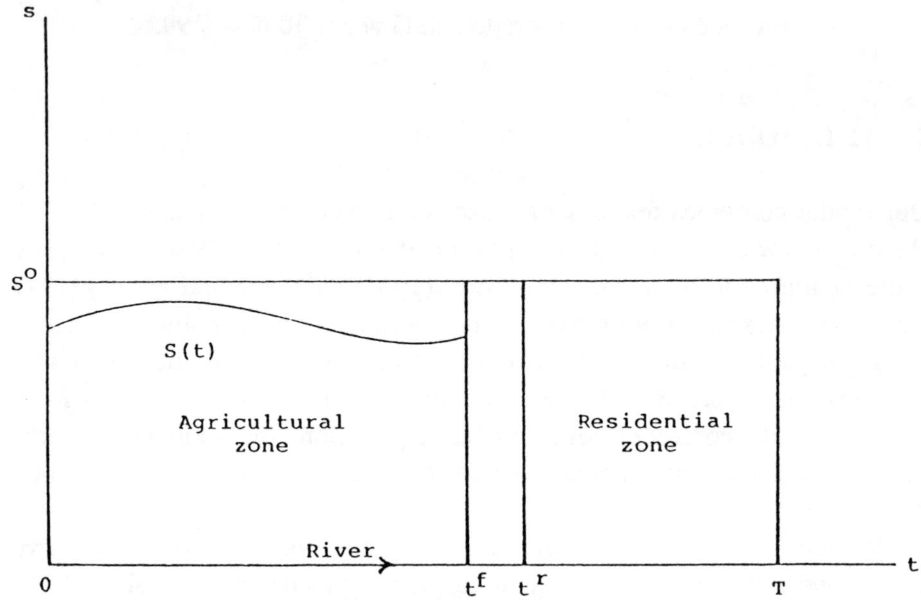

Fig. 1. Regional Land Use

the set of "active" farming locations is a subset of $[0, T]X[0, S^o]$, the possible farming locations. Output is produced according to the production function

$$y(t, s) = f(x(t, s)), \tag{1}$$

where $f(.)$ is assumed to be increasing and strictly concave. Aggregate production is

$$\int_0^{t^f} \int_0^{s(t)} y(t, s) dt\, ds \equiv Y^s \tag{2}$$

More generally, f could be assumed to be μ-measurable, where μ is Lebesque measure (see Mäler (1974) for use of such an approach). However, little generality is lost and much simplicity of exposition is gained by assuming that the set of active farms is such that we can employ ordinary, Riemann integrals.

The input x is purchased in a national market at a fixed price of q per unit. We assume that x can be transported costlessly within the region. Agricultural output is sold at the central business district (CBD) where a price of $p(T)$ is received per unit. Following Solow and Vickery (1971) and Hochman *et al.* (1977), we assume that the agricultural goods is costless to transport in the s-dimension, but is costly to transport in the t-dimension. Let $p(t)$ be the price net of transport costs received at t, with $p'(t) > 0$ and $p(0) \geq 0$. Note that we ignore important aspects of the rural labor market; farmers supply labor perfectly inelastically only to their own farm and we have suppressed this input in the notation.

Farming generates pollution according to the function $z(x(t, s))$. In this specification there is no way to reduce pollution other than to reduce use of the only agricultural input. Other ameliorating inputs could be envisioned (as in Hochman *et al.*, 1977) and are important in practice. Modification of our model to incorporate them is straightforward, but would yield no additional insights into the issue we address. Farmers must pay a pollution tax (the SPT) of $\tau(t, s)$ per unit of emissions from the farm field, as well as a lump-sum tax or subsidy on farming at location t of $K(t, s)$. Farmers also have to pay a rental fee to a landlord of $R(t, s)$.

Farmers are assumed to farm as if they maximize profits, given by

$$p(t)f(x(t,s)) - qx(t,s) - \tau(t,s)z(x(t,s)) - K(t,s) - R(t,s) \tag{3}$$

The first order necessary condition for profit maximizing input use by active farms is

$$p(t)f'(.) = q + \tau(.)z'(.) \tag{4}$$

We assume that equation (4) has a unique solution $x^*(p(t), q, \tau(t, s))$. Let the (indirect) profit function be denoted by $\pi^*(p(t), q, \tau(t, s), K(t, s), R(t, s))$. It is readily verified that $x^*(.)$ is decreasing in τ. While in the absence of pollution $x^*(.)$ is increasing in t (i.e. farmers closer to the market farm more intensively), this cannot be concluded once spatial pollution taxes are considered.

To determine the set of active producers in the long-run equilibrium, given the array of prices and taxes, entry will occur until zero profits are made at all active locations, or until a capacity constraint is reached at a location in the s-dimension, in which case positive profits could be made. Naturally, the equilibrium price vector will be determined such that markets clear, zero profits are made at all active locations, and rents absorb any potential profits at capacity-constrained locations.

Manufacturing

Manufacturing is conducted by a variable number of identical, competitive firms. The only input to production is labor, which is purchased from consumers at a wage rate of w per unit. Let their be N of such firms, and let $f^m(L)$ be the production function of the representative firm. Output is sold at a price of p^m. The profits of the representative firm are given by

$$p^m f^M(L) - wL \tag{5}$$

We assume profit maximization in manufacturing. Let $L^*(p^m, w)$ be the labor demand function and $\pi^m(p^m, w)$ be the profit function. $L^*(.)$ is increasing in p^m and decreasing in w.

We assume costless entry and exit in the manufacturing sector, as well as constant returns to scale. Hence, zero profits are made in the long run equilibrium and we need not be concerned with ownership of firm profits.

Pollution

Agricultural production at (t, s) generates pollutants according to the function $z(x(t, s))$, with $z'(.) > 0$. The impact of these pollutants is assumed not to be altered by movements from the farm field to the river. Thus, we may drop the s argument from the emission tax τ, lump sum tax K, and rental payment R, since $f(.)$ and prices do not vary in the s-dimension either.

In order to distinguish between the interventions proposed by Henderson (1977) and by Carlton and Loury (1980, 1986), we specify that the effect of agricultural production at t on river water quality at t depends both on the emissions of pollution per firm and on the number of firms at t. This is the general form of the damage function studied by Carlton and Loury. This is a controversial assumption. Spulber holds that only emissions cause damages, while Carlton and Loury (1986) respond that the distribution of emissions across firms can matter as well. We do not take a stand on this issue, and merely seek to understand the influence of assumptions regarding the form of the damage function on efficient regulatory policy. However, any reasoning of why the distribution of emissions across firms may matter which depends on spatial considerations is ruled out here, since space does not influence pollution impacts in the s-dimension in our model. In our model the "number of farms" at t is identified by the length of the interval $[0, S(t)]$ (Mäler, 1974). Hence, to capture the Carlton-Loury notion we let $g(z(x(t)), S(t))$ determine the effect of farming in $[0, S(t)]$ on river water quality at t.

As pollutants move downstream towards the lake, their impact on water quality is ameliorated, so that in the absence of new additions of pollutants, river water quality improves as t increases. Let $Z(t)$ be a state variable representing river water quality at t. The transition equation for $Z(t)$ is

$$\dot{Z}(t) = G(g(z(t), s(t), Z'(t), t) \tag{6}$$

In the model investigated by Henderson (1977), the pollution disamenity suffered by a consumer depends on the pollution accumulated at the boundary between the polluting sector and the residential area, as well as the distance between the consumer's residence and that boundary. An analogous formulation here specifies that lake water quality depends on accumulated impacts at the boundary of the agricultural area (i.e. $Z(t^f)$ and the distance between that boundary and the lake. Thus, parallel to Henderson, we assume that lake water quality, Q, is determined by a function

$$Q = \Gamma(Z(t^f), t^f, T) \tag{7}$$

(compare Henderson eq. [3]).

A natural alternative to (7) is to write $Q = \Gamma(Z(T))$ and to use the transition equation (6) with $g = 0$ to translate $Z(t^f)$ into $Z(T)$. Since the position of the boundary of the agricultural area is arbitrary from a biophysical standpoint, as long as pollution effects are governed by equation (6) for t everywhere in the interval $[0, T]$, it is reasonable to construe (7) as equivalent to $\Gamma(Z(T))$ and to concentrate on the implications of alternative forms of the transition equation $G(.)$. It might also be supposed that residential land uses also cause surface water pollution, for example from fertilization of and pesticide applications to lawns. This complication is not central to our analysis, but could be incorporated with fairly obvious modifications.

Consumers

Consumers use income from wages, agricultural land rents, and tax rebates from the government to purchase a consumption bundle consisting of the manufactured goods, the agricultural goods, and housing services. The consumers endowment of labor, equal to one unit, is perfectly inelastically supplied. Consumers commute to the CBD once each period, the cost of which depends on how far they live from the center. Also included in the bundle of goods they consume is lake water quality $Q(T)$. Although consumers might be affected by river water quality as well as lake water quality, this complication is not studied here. Hochman *et al.* (1977) analyze of this issue in their paper, but do not study its effects on an optimal boundary between the agricultural and residential areas. This generalization is not trivial. If consumers are affected by stream water quality it is plausible that there should be a gap between the two areas; here we study the factors determining the optimal position of a common boundary. If a gap is admitted, then much more complex mathematical issues arise, since one then has a control problem with optimal "start-up" times. While techniques exist for the study of such problems, this analysis is best left for future research.

Let Y denote consumption of the agricultural good, Y^m be consumption of the manufactured good, $h(t)$ be housing services consumed at t, $r(t)$ be residential rents at t, and $c(t)$ be commuting costs, with $c'(t) < 0$. We assume that consumers' preferences can be represented by a utility function $U(.)$, which is increasing in all its arguments and quasi-concave. We assume too that consumption choices are made as if they solve the problem

$$\max U(Y, Y^m, h, Q)$$

$$\text{s.t.} \tag{8}$$

$$p(T)Y + p^m Y^m + r(t)h \leq w - c(t) + R^c + T^c,$$

where R^c and T^c are the representation consumer's share of the agricultural rents and tax receipts, respectively. The entire analysis is greatly simplified if it is assumed that consumers share equally in aggregate rents and tax rebates. Since the concerns that arise if this assumption is dropped are not central to this paper, the equal-share assumption is imposed.

We denote demands by $Y^*(.), Y^{m*}(.)$ and $h^*(.)$; these are functions of the price vector $(p, p^m, r(t))$, net income $m(t)$, which is defined by the right-hand side of the constraint in (8), and lake water quality Q. The indirect utility function is denoted by $V(p, p^m, r(t), m(t), Q)$, or by $V(t)$ where convenient. As usual, demands can be related to derivatives of the indirect utility function via Roy's identity. We also have occasion to work with the marginal willingness to pay for lake water quality denoted by $\phi^*(p, p^m, r(t), m(t), Q)$; the function $\phi^*(.)$ equals the derivative of the indirect utility function with respect to Q, divided by the marginal utility of income (Mäler, 1974).

Aggregate demands and labor supply are obtained by integrating individual demands over the residential area, which falls in the interval $[t^r, T]$. To put demands on a per unit area basis, they are divided by $h^*(t)$. Due to the assumption that one unit of labor is supplied each period, aggregate labor supply and the regional population of consumers are the same. Letting a superscript "a" denote aggregate demands and supplies we have

$$Y^d = \int_{t^r}^{t} \int_{0}^{s(t)} Y^*(.)/h^*(.) ds\, dt \tag{9}$$

(agricultural good demand)

$$Y^{md} = \int_{t^r}^{t} \int_{0}^{s(t)} y^{m^*}(.)/h^*(.) ds\, dt \tag{10}$$

(manufactured good demand)

$$L^s = \int_{t^r}^{t} \int_{0}^{s(t)} 1/h^*(.) ds\, dt \tag{11}$$

(labor supply)

We now turn to the equilibrium of this model.

3. COMPETITIVE EQUILIBRIUM

As usual, a competitive equilibrium is defined as a set of prices such that, at these prices the demands and supplies defined above, plus any quantities imported or exported from the region, are equal; the balance of trade is zero; farmers and manufacturers make zero profits; consumers have no incentive to re-locate; and there are no opportunities for profits to be made from changes in land use. We have not shown that a competitive equilibrium exists for our economy; however, existence has been demonstrated for a very similar setup, under assumptions of Cobb-Douglas forms for production and utility functions, by Miyao *et al.* (1980).

To simplify our analysis, we assume that the region is small and that it is open. By small is meant that the region takes the prices of traded goods as given. We assume that the agricultural and manufactured goods are traded; hence, $p(T)$ and p^m are exogenous. We already have assumed that the price of the agricultural input, q, is exogenous, as are the costs of transportation of agricultural goods and of commuting. Taking the taxes levied by the government as given as well, the endogenous variables determined by the equilibrium conditions discussed above are the agricultural and residential land rent schedules $R(t)$ and $r(t)$, the wage rate w, water quality, Q, the residential and agricultural areas $[0, t^f]$ and $[t^r, T]$, and the regional demands, supplies and traded quantities.

Rather than fully investigate the competitive equilibrium, which is fairly standard, we only mention some of its properties, focusing on land use determinations. Letting superscript T represent imported quantities, the full set of equilibrium conditions is given by:

$$L^*(p^m, w) = \int_{t^r}^{T} \int_{0}^{s(t)} 1/h^*(.)ds\, dt \tag{12}$$

(labor market clearing);

$$Y^m(.) + Y^{mT} = \int_{t^r}^{T} \int_{0}^{s(t)} y^{m*}(.)/h^*(.)ds\, dt \tag{13}$$

(manufactured good market clearing);

$$Y^*(.) + Y^T = \int_{t^r}^{T} \int_{0}^{s(t)} y^*(.)/h^*(.)ds\, dt \tag{14}$$

(agricultural market clearing);

$$p^m Y^{mT} + p(T)Y^T = 0 \tag{15}$$

(zero balance of trade);

$$V(p(T), P^m, r(t), m(t), Q) = V^* \tag{16}$$

(no migration);

$$\pi^{m*}(p^m, w) = 0 \tag{17}$$

(no entry in manufacturing);

$$R(t) = p(t)f(x^*(.))qx^*(.) - \tau(t)z(x^*(.)) - K, all\ t \tag{18}$$

(no entry in agriculture);

$$R(t^c) = r(t^c) \tag{19}$$

(no arbitrage in land use).

It is clear that in the absence of pollution taxation, the agricultural rent gradient is upward-sloping, since the net price $p(t)$ is rising as one gets closer to the CBD located at T. Formally, this follows from the envelope theorem, which can be used to show that the derivative with respect to t of the profit function of a representative farm at t equals the derivative of the price function times the quantity supplied at t. Depending on how the emission tax is levied, this result need not hold in the regulated equilibrium. Below we show that the Pigouvian tax falls out of our setting as one gets farther from the CBD; perhaps this more than offsets the decrease in the net output price, giving rise to rents that increase with distance (see Henderson, 1977). For convenience, we assume that the after-tax agricultural rent gradient falls as distance from the CBD increases.

It is easily shown using the envelope theorem that the residential rent gradient falls as commuting distance increases. As usual in such models, we take it that the residential rents at the CBD are higher than agricultural rents there, and that they have a steeper slope than agricultural rents.

Under these assumptions, the no-arbitrage equilibrium condition regarding land use changes dictates that the residential section lies close to town with the agricultural area lying beyond it. There is a single "competitive" boundary, t^c such that $t^f = t^c = t^r$, and, importantly, the residential rents equal the agricultural rents at this competitively-determined boundary. The land use pattern in the competitive equilibrium for our specification is depicted in Figure 2.

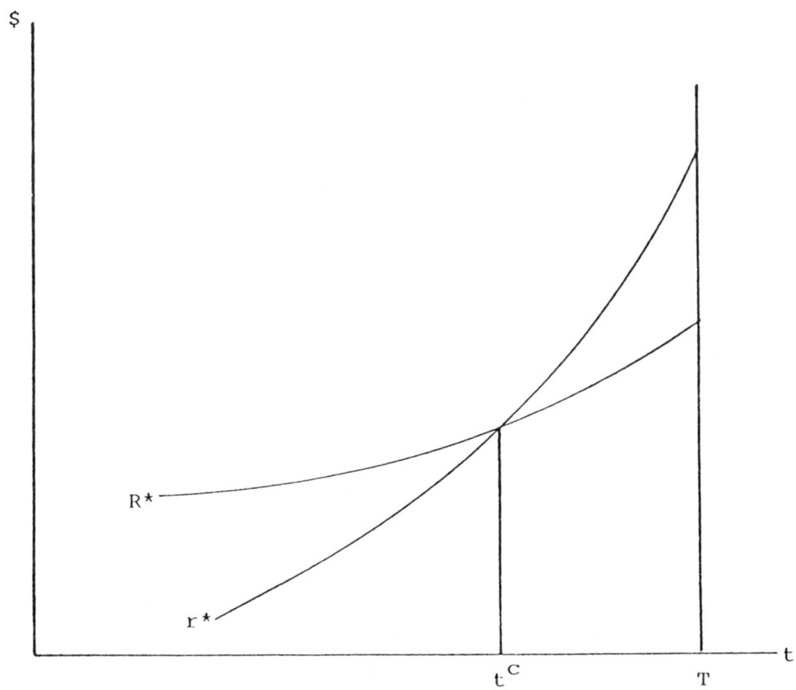

Fig. 2. Competitive Land Allocation

4. THE PARETO OPTIMUM

In this section we study the optimal configuration of land uses in the region. We are not interested here in a complete characterization of the optimal allocation, but wish to focus our attention on its spatial aspects. For a more complete derivation of the optimal allocation in a very similar setting, see the appendix in Henderson (1977). Three elements of the overall allocation will be central to our analysis: the allocation of the agricultural input $x(t)$ over t, the optimal width of the agricultural zone $S(t)$ at each t, and the position of the boundary between the residential and agricultural zones. We follow Hochman *et al.* (1977) and specify our problem as one of optimal control.

We can characterize a Pareto optimum by maximizing same social welfare function, e.g. an increasing function of the indirect utilities. Since indirect utility is fixed at V^* by assumption, and residential rents adjust to any changes in the system so as to maintain this inequality, it is clear that a suitable social welfare function for our purposes is aggregate residential rents. We assume that V(.) is twice differentiable and that $\partial V/\partial r < 0$; hence, we can conclude that residential rents are a differentiable function $r^*(\theta(t))$ where $\theta(t) = (P(T)p^m, m(t), Q, V^*)$. We will assume that all available residential land is filled in the interval $[t^r, T]$.

Note that by working with the indirect utility function we have assumed that all consumer choices adjust optimally to allocations in the farm sector, and they have therefore been "maximized out." In essence, we have a two-staged control problem: one stage for the agricultural sector and one for the residential sector. The latter control problem has been solved and the indirect objective function added to the former problem as a "scrap value" function.

Our objective function is, therefore,

$$J = \int_{t^0}^{T} \int_{0}^{S} r^*(.)ds\, dt$$

We seek to maximize J by choosing the control vector $u(t) = (x(t), S(t))$ and the free endpoint of the system, t^f. We assume that the residential area has positive measure so that we can take as non-binding the endpoint constraint $t^f \leq T$. The maximization takes place subject to the constraints governing how residential income is affected by the evolution of net farm income (agricultural rents) over space, the induced changes in water quality, the land use constraint $S(t) \leq S^0$, and the technical and other relationships previously specified. Our analysis is based on the function expressing residential rents as a function of income and water quality. Income is "accumulated" through space through agricultural rents. We may focus on pretax farm profits since taxes received by the government are rebated to consumers according to the same distribution as agricultural rents. Thus, they are subtracted from rents but added back to income and therefore net out of the analysis.

One complicating factor arises from the presence of wage income in the income term. Wages are determined in equilibrium by the market-clearing equation (12). Hence, the wage is a function of regional population, which is in turn a function of population density $(1/h^*(.))$, the agricultural-residential boundary, and water quality. A full analysis would consider these general equilibrium effects in the determination of an optimum. This consideration introduces details that are not central to our analysis here; hence, we assume that wages do not respond to alterations in farm-level choices. Our analysis is a restricted one, but it provides a basis for a more general policy analysis in future research.

Formally, the social planner's problem can be written

$$\max J$$
$$\text{s.t.}$$
$$\dot{I}(t) = \int_0^{s(t)} pf(x(s,t)) - qx(s,t)ds$$
$$\dot{Z}(t) = G(g(z_t, S_t), Z_t, t)$$

This is a straightforward optimal control problem. The objective function, aggregate residential rents, depends on the boundary between the residential and agricultural sectors via the water quality function in equation (7). However, the control problem in the agricultural sector does not depend on this boundary, since the Hamiltonian consists solely of transition equations which do not depend on the boundary. The results of Tomiyama and Rossana (1989, Theorem 1) can be applied to show that the above can be solved by an application of the maximum principle to the problem in the agricultural zone, where the residential rents act as a "scrap value" function. If the Hamiltonian for the agricultural region also depended on the boundary, the more elaborate "matching" conditions of Tomiyama and Rossana would apply. Naturally, there are a number of relevant models in which such a dependence would occur; for example if input prices on the farm depended on regional wage levels. However, these more general effects do not appear in our model, and the simpler analysis applies. The Hamiltonian for our problem is

$$\mathcal{H} = \lambda(t)\dot{I}(t) + \gamma(t)\dot{Z}(t)$$

where $\lambda(t)$ and $r(t)$ are costate variables. The Lagrangean function for the constrained maximization of the Hamiltonian is

$$\mathcal{L} = \mathcal{H} + \beta_t[S^0 - S(t)]$$

The first order necessary conditions for a maximum include:

$$0 = \frac{\partial \mathcal{L}}{\partial x(s,t)} = \lambda_t \int_0^{S(t)} [p(t)f'(x) - q]\, ds + \gamma_t \left[\frac{\partial G}{\partial g}\frac{\partial g}{\partial z} z'(x)\right] \tag{20}$$

$$0 = \frac{\partial \mathcal{L}}{\partial s(t)} = \lambda_t \left[p_{(t)}(x) - qx\right] + \gamma_t \left[\frac{\partial G}{\partial g}\frac{\partial g}{\partial S}\right] - \beta_t \tag{21}$$

$$\beta_t \geq 0;\ S^0 - S(t) \geq 0;\ \beta_t \left[S^0 - S(t)\right] = 0$$

$$\dot{\lambda}_t = -\frac{\partial \mathcal{H}}{\partial I_t} = 0 \tag{22}$$

$$\dot{\gamma}_t = \frac{-\partial \mathcal{H}}{\partial I_t} = -\frac{\partial G}{\partial Z}(.)\gamma_t \tag{23}$$

Transversality conditions for the costate variables are (Tomiyama and Rossana, 1988)

$$\lambda(t^0) = \frac{\partial J}{\partial I(t^0)} = \frac{1}{N*}\int_{t^0}^{T}\int_0^{s^0} \frac{\partial r^*}{\partial M}(.)ds\, dt \tag{24}$$

$$\gamma(t^0) = \frac{\partial J}{\partial Z(t^0)} = \int_{t^0}^{T}\int_0^{S^0} \frac{\partial r^*}{\partial Q}(.)\frac{\partial \Gamma}{\partial Z(t^0)} ds\, dt \tag{25}$$

$$\mathcal{H}(t^0) + \frac{\partial J}{\partial t^0} = 0 \tag{26}$$

We begin our analysis of the solution with (24). Applying the implicit function theorem to the equation defining $r^*[i.e. V(., r^*(.)) - V \equiv 0]$ we have

$$\frac{\partial r^*}{\partial M} = \frac{-\partial V/\partial M}{\partial V/\partial r} = \frac{-1}{h^*(.)}, \tag{27}$$

by Roy's Identity. Thus, the double integral on the rhs of (24) equals the regional population, and we can conclude that $\lambda(t^0) = 1$. From (22), we have $\lambda(t) = 1$ for all t.

Solving equation (23) provides

$$\gamma(t) = \gamma(t^0) \exp\left[\int_t^{t^0} \frac{\partial G}{\partial z}(z)dz\right] \tag{28}$$

Again using the Implicit Function Theorem and Roy's Identity we discover that

$$\frac{\partial r^*}{\partial Q} = \frac{\partial v/\partial Q}{\partial v/\partial M}(.)\frac{1}{h^*(.)} \tag{29}$$

Substituting this expression into (25) and the result into (28) we obtain

$$\gamma(t) = \frac{\partial \Gamma}{\partial Z(t^0)} \int_{t^0}^{T} \int_0^S \frac{\partial v/\partial Q}{\partial v/\partial M} \frac{1}{h^*(.)} ds dt \exp\left[\int_t^{t^0} \frac{\partial G}{\partial z}(z)dz\right] \tag{30}$$

Equation (30) reveals that the shadow value of pollution accumulation is the product of three terms. The first gives the impact on lake water quality of a small increase in pollution accumulated at the boundary between the residential and agricultural zones. Of course, from the solution to the transition equation this term is just $\Gamma'(\theta(T))$. The second term is aggregate marginal willingness to pay for improvements in lake water quality. This is the aggregate of ϕ^*. The third term is the spatial term. It translates marginal willingness to pay, defined as of the boundary t^0, to an effect at t. This term accounts for the degradation of the impact of pollutants as they move between t and t^0.

So far the analysis is entirely standard. We turn our attention now to the two remaining issues: determining $S(t)$, the land used in agriculture at each t, and the position of the optimal boundary between the residential and agricultural zones.

Regarding $S(t)$, the condition for an optimum is given by

$$0 = \frac{\partial \mathcal{L}}{\partial S(t)} = \lambda_t \left[p(t)f(x_t) - qx_t\right] + \gamma_t \frac{\partial G}{\partial g}\frac{\partial g}{\partial S} - \beta.$$

$$\beta_t \geq 0; S^0 - S(t) = 0; \beta_t \left[S^0 - S^0 - S(t)\right] = 0 \tag{31}$$

Recalling that $\lambda_t = 1$, this specifies that pre-tax farm profits, plus the shadow cost of additional farmland on water quality, should equal the shadow value of relaxing the land availability constraint. Naturally, this later term, β_t, is zero if land is abundant at t, in which case profits plus pollution costs equal zero. Equation (31) says that the marginal value of land in agriculture is profit net of pollution costs, as one would expect. For the boundary between the residential and agricultural zones, we use the endpoint condition in (26). Written out, this becomes [using (29)]

$$\int_0^{s(t^0)} \left[p(t^0)f(x(t^0)) - qx(t^0)\right] ds + \gamma(t^0)G(g(z(x(t^0)), S(t^0)), Z(t^0), t^0)$$
$$+ \frac{\partial \Gamma}{\partial t^0} \int_{t^0}^{T} \int_0^{S} \frac{\partial V/\partial Q}{\partial V/\partial M} \frac{1}{h^*(.)} ds dt - \int_0^{S^0} r^*(t^0, .) ds = 0, \qquad (32)$$

where the last two terms correspond to $\frac{\partial J}{\partial t^0}$.

Equation (32) is a central result of this paper. The first term is the increment to regional income from adding a slice of farmland at the boundary. The second term is the shadow cost of adding this farmland in terms of water quality. The third term is aggregate marginal willingness to pay to have the agricultural boundary moved closer to the lake where water quality is demanded; this term is negative. The last term is the value of residential land lost to the region as the agricultural zone is extended marginally. Thus, (32) requires that marginal benefits and marginal costs of expanding the agricultural zone be equated.

5. OPTIMAL SPATIAL ENVIRONMENTAL POLICY

On this section we investigate into the form of environmental policies which can be used to sustain an optimum as a competitive equilibrium. First, we consider input use in farming. It is clear how the shadow value of pollution, as given in equation (30), relates to the proper Pigovian tax on emissions. Using the result that $\lambda_t = 1$, (20) is identical to the farmer's first order necessary condition for profit maximization in equation (4) if the SPT is set according to

$$\tau(t) = \frac{-\gamma(t)}{S(t)} \frac{\partial G}{\partial g} \frac{\partial g}{\partial z} \qquad (33)$$

where $\gamma(t)$ is given by (30). That is, the appropriate SPT is the shadow cost of pollution in the stream at t, times the marginal impact of emissions at the farm field on river pollution at t. Substitution from (30) into (33) yields a more complete description of the optimal emission tax.

Next, we consider the optimal amount of land to allocate to agriculture at each distance t. Using the expression in (33) as the SPT on emissions, we can rewrite (21) as

$$p(t)f(x_t) - qx_t\tau_t z(x_t) - \frac{\gamma_t}{S_t}\frac{\partial g}{\partial g}\left[-\frac{\partial g}{\partial z}z(x_t) - \frac{\partial g}{\partial S}S_t\right] = \beta_t \qquad (34)$$

The first three terms on the left hand side q (34) are the farm profits at t net of the SPT. If a tax or subsidy $K(t)$ we placed on farming such that

$$K(t) = \frac{\gamma_t}{s_t}\frac{\partial G}{\partial g}\left[\frac{\partial g}{\partial z}z(x_t) - \frac{\partial g}{\partial S}S_t\right] \qquad (35)$$

then equation (34) would state that after-tax farm profits equal the shadow value of agricultural land. Naturally, this result can be decentralized if we identify β, with agricultural land rents. In this case, the complementary slackness conditions on the land constraint make sense. If land is not fully utilized at t, then rents are zero and after tax profits are driven to zero as well. However, if the available land at t is fully employed then it commands a positive rent.

The tax specified by (35) is exactly the lump sum tax/subsidy scheme identified by Carlton and Loury (1986), adapted to our model. If it is levied on farming along with the SPT in (33), then both the efficient intensity of farming (application of x_t and efficient number of farms (S_t) is achieved in the decentralized equilibrium. This reveals a problem with the Carlton and Loury analysis. It involves a rebate of the emission tax paid, plus a term related to the marginal pollution impacts of adding polluting farms at t. It seems implausible that this would be treated as a "lump sum" tax or subsidy by farms, since it obviously is related to production decisions via the emission tax rebate.

In any event, a sufficient condition under which the Carlton-Loury tax is zero is easily identified. If pollution effects depend only on aggregate output of emissions at t, then the tax is zero at least if we are not at the agricultural/residential boundary. Formally we have

Theorem 1 *If* $g(z(x_t), S_t) = \int_0^{S_t} z(x_t)ds = z(x_t)S_t$, *then* $K(t) = 0$ *for* $t \neq t^0$, *and* $\tau(t) = -\gamma(t)\partial G/\partial \tilde{z}(t)$, *where* $\tilde{z}(t) = \int_0^{s(t)} z(t)ds$.

Proof: In this case we have that $\partial g/\partial z = S$ and $\partial g/\partial S = z$. Then clearly the expression in brackets in (35) is zero. Using these results in (33) provides the result, since the conditions for the efficient input use in agriculture and efficient number of firms become

$$p_t'(x_t) - q + \gamma_t\frac{\partial G}{\partial \tilde{z}}z'(x_t) = 0 \qquad (36)$$

and

$$p_t f(x_t) - qx_t + \gamma_t \frac{\partial G}{\partial z} z(x_t) = \beta \tag{37}$$

If an emission tax is imposed equal to $-\gamma_t \frac{\partial G}{\partial z}$, then (36) is the first order condition for a farmer's profit maximization problem and (37) is the condition stating that after tax profits equal land rents.

Theorem 1 is exactly the result of Carlton and Loury (1986) and Spulber (1985) regarding the efficiency of Pigouvian taxation in the long run in a non-spatial model.

Now, we turn our attention to the determination of the optimal boundary between the agricultural and residential zones. We are interested in the ability to "decentralize" the social planner's conditions for an optimum as stated in equation (32).

Let Φ be aggregate marginal willingness to pay to avoid increases in lake pollution. That is, Φ is the integral over residential households of $\phi^*(.)/h^*(.)$. Since in an optimum pre-tax farm profits and residential rents at any t are fixed, we can write the first term of (32) as $[pf^*(t^o) - qx^*(t^o)]S^*(t^o)$ and the last term as $r^*(t^o)S^o$ (recall the earlier assumption that residential land fills the available space at each t). We impose now an assumption that $S^*(t^o) = S^o$, so that agricultural rents are taken to be positive at the boundary. Evaluating equation (30) at $t = t^o$ and substituting into (32) yields

$$[pf^*(t^o) - qx^*(t^o)]\, S^o + \Phi \left[\frac{\partial \Gamma}{\partial Z(t^o)} G(g(t^o), t^o) + \frac{\partial \Gamma}{\partial t^o} \right] = r^*(t^o)S^o \tag{38}$$

Adding and subtracting tax payments, using (33) for the SPT on emissions and (35) for K yields

$$\pi^*(t^o) + \frac{\Phi}{S^o} \left[\frac{\partial \Gamma}{\partial Z(t^o)} \left\{ G(t^o) - \frac{\partial g}{\partial S} S^o \right\} + \frac{\partial \Gamma}{\partial t^o} \right] = r^*(t^o) \tag{39}$$

Naturally, from the equilibrium condition in equation (19), agricultural and residential rents are equated at the boundary. Thus, from the definition of agricultural rents, we have the condition

$$\pi^*(t^c) = r^*(t^c) \tag{40}$$

Comparing (39) and (40), it is clear that imposition of SPTs and the Carlton-Loury tax K is sufficient for a decentralized equilibrium to sustain a Pareto optimum only if (since rents are continuous under previous assumptions)

$$\frac{\partial\Gamma}{\partial Z(t^o)}\left\{G(t^o)-\frac{\partial g}{\partial S}S^o\right\}+\frac{\partial\Gamma}{\partial T_0}=0 \tag{41}$$

It is not possible to determine the sign of (41) in general. Thus, in the absence of the "additional intervention" identified by Henderson (1977), the optimal boundary might be farther or closer to the lake than is the decentralized boundary. In either event, a zoning tax must be placed on agriculture (or equivalently, a subsidy to residential land) in the area between t^c and t^o. The appropriate zoning tax is given by

$$K^b=\frac{\Phi}{S^o}\left[\frac{\partial\Gamma}{\partial Z(t^o)}\left\{G(t^o)-\frac{\partial g}{\partial S}S^o\right\}+\frac{\partial\Gamma}{\partial T^o}\right] \tag{42}$$

It is important to note that, even if the postulate of Theorem 1 hold, so that the tax K in (35) is zero, it still generally will be the case that SPTs alone do not suffice to ensure efficiency. Under the supposition that the optimal zoning tax on agriculture is positive, the regional configuration is depicted in Figure 3.

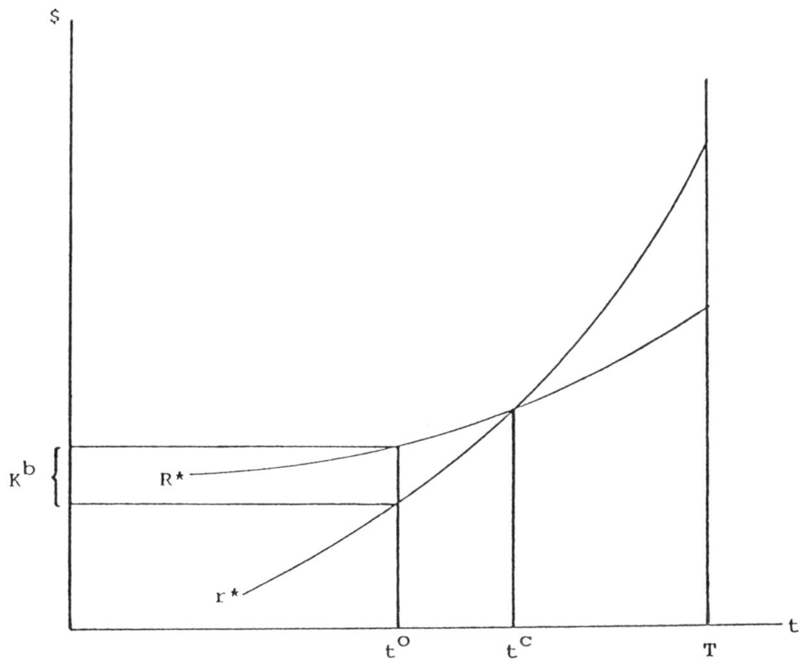

Fig. 3. Optimal Land Allocation

The remainder of our attention in this section is directed to establishing a sufficient condition under which (41) holds, and $K^b = 0$. Note first that if the postulate of Theorem 1 is invoked, so that $g(z, S) = z^* S^*$, the second term in braces in (41) is just $z^*(t^o)S^o$. Suppose next that $G(g, Z, t)$ is linear in g and so can be written

$$G(g, Z, t) = g - \xi(Z_t, t) \tag{43}$$

Then (41) becomes

$$-\frac{\partial \Gamma}{\partial Z(t^o)}\xi(Z_t, t) + \frac{\partial \Gamma}{\partial t^o} \tag{44}$$

According to (44), whether the Henderson intervention is necessary hinges on how solutions to the transition equation respond to with changes in the pollution state variable, $Z(t)$, at the boundary, versus a change in the boundary itself. It is straightforward to verify that, if the term $\xi(Z_t, t) = \beta(t)Z_t$ then (41) is zero. We state all of this as

Theorem 2 *If $G(g(z, S), Z, t) = z_t S_t - \beta(t)Z_t$, then the SPTs alone suffice to ensure efficiency.*

Proof: Since $z(t) = 0$ for all $t \in [t^o, T]$, we can solve the transition equation to find

$$Z(T) = Z(t^o)\exp\left[\int_{t^o}^{T} \beta(t)dt\right] \tag{45}$$

Differentiation shows that (44) is zero.

The form of the transition function assumed in Theorem 2 is the same as that employed by Hochman *et al.* (1977) in their paper on spatial pollution They further impose that $\beta(t)$ is a constant, but his is immaterial to our considerations. As mentioned in the introduction, Hochman *et al.* did not examine the issue of the optimal boundary between agricultural and residential areas, but implicitly asserted that decentralized determination of this boundary, given SPTs, would generate an optimal configuration of land uses in their region. In fact, this is the case. However, this is not a general result, as we have seen.

The conditions of Theorem 2 may perhaps be relaxed. We will not attempt a complete characterization here, since in any applied situation, one would be given a transition equation and one could check to see if the spatial tax should be imposed or not. The central result of this paper is that one does, in fact, need to check, and that the results of Henderson and of other authors can be reconciled by examining the transition equation on pollutants.

6. SPATIAL ISSUES IN NONPOINT SOURCE POLLUTION

In this section of the paper, we very briefly discuss some implications of the above analysis for policy to control nonpoint source pollution (NPS). The key distinction between the point source approach taken above and the nonpoint source setting regards uncertainty and information. Obviously, generalizing the above framework to incorporate uncertainty and asymmetric information is a formidable task, the object of current research (see Moretto and Graham-Tomasi, 1992). However, we can offer some observations on existing incentive schemes and how the above model might be generalized.

First, it seems apparent that any simple tax scheme that does not achieve marginal budget balance will not be efficient, if it is not accompanied by spatial interventions to restore the optimal boundary or careful attention given as to how lump sum taxes/subsidies are collected. By budget balance is meant that the sum of collections of tax revenues equals the sum of marginal damages over consumers.

Thus, the incentive scheme proposed by Segerson (1988), which does not lead to budget-balance, might not achieve long run efficiency. In the Segerson scheme, the regulator collects the full value of marginal damages from each polluter, rather than just their share. In this case, tax collections exceed damages and, relative to the tax collections we specify, too much money is extracted from farmers, the agricultural rent gradient is too low, the residential gradient is too high, the boundary between the sectors is too far from the lake, and the lake is insufficiently polluted. Note that this does not depend on any particular form of the transition equation; this spatial inefficiency holds even if the transition equation satisfies the conditions of Theorem 2. That said, it also is clear that efficiency can be achieved by lump sum interventions which truly are lump sum, i.e. applied to both the agricultural and non-agricultural sectors, and other regions as well, in a manner such that entry-exit decisions on particular plots are not distorted from the type of efficiency condition we identify here.

In a subsequent analysis by Miceli and Segerson (1991) of joint liability in tort law, which bears a close resemblance to the NPS problem studied in Segerson, firms are given a "credit" from their tax payments, equal to aggregate damages without the firm operating. Adding this credit to the incentive scheme may or may not lead to budget balance, but it does lead (in their model) to correct long run incentives. Overall budget balance depends on possible interactions among firms in causing damages. It is difficult to assess the implications within our framework of any non-budget balance that might arise in the Miceli-Segerson model. Are there any general equilibrium effects of non-budget balance that would arise in our model that are not identified in theirs? How does one capture the interdependencies among polluters that give rise to an absence of budget balance in terms of a transition equation? How does one specify the "credit" scheme when there is a coninuum of polluters? These are important topics for further research.

Regarding the issue of incorporating NPS considerations into our framework, it would seem that making the transition function a stochastic differential equation would incorporate many of these effects. That is, given a regulatory policy, the regulator, because he/she cannot observe type and/or actions within each "slice" of the region in the s-dimension, is faced with an uncertain increment to Z_t. The crux of the problem, of course, is specifying the distribution of types across space in both dimensions in a manner that is both "realistic" and tractable.

ACKNOWLEDGEMENTS

Much of this research was conducted while the authors were Associate Professor and Graduate Research Assistant, Department of Applied Economics, University of Minnesota. They thank the Minnesota Agricultural Experiment Station for financial support. More recent efforts have been supported in part by the Fondazione ENI, Milan, Italy and in part by the U.S. Agency for International Development, Environmental and Natural Resource Policy and Training Project (EPAT). An earlier version was presented at the Workshop on Nonpoint-source Pollution sponsored by the Fondazione ENI in Milan, Italy, in April, 1992. Helpful comments have been provided by Jim Opaluch, Kathy Segerson, Michele Moretto, Cesare Dosi, Jim Oehmke and Jacque Thisse. Related material was presented in a paper at the European Association of Environmental and Resource Economists meetings in Venice, Italy, in May, 1990.

REFERENCES

Carlton, D.W., Loury, G.C. (1980). "The Limitation of Pigouvian Taxes as a Long-Run Remedy for Externalities", *The Quarterly Journal of Economics*, XCV (November), pp. 559–566.

Carlton, D.W., Loury, G.C. (1986). "The Limitation of Pigouvian Taxes as a Long-Run Remedy for Externalities: An Extension of Results", *The Quarterly Journal of Economics*, CI (August), pp. 631–634.

Freeman, A.M. (1984). "Depletable Externalities and Pigouvian Taxation", *Journal of Environmental Economics and Management*, II, pp. 173–179.

Henderson, J.V. (1977). "Externalities in a Spatial Context: The case of air pollution", *Journal of Public Economics*, 7, pp. 89–110.

Hochman, E., Pines, D., Zilberman, D. (1977). "The Effects of Pollution Taxation on the Pattern of Resource Allocation: The Downstream Diffusion Case", *The Quarterly Journal of Economics*, XCI (November), pp. 625–638.

Laffont, J.J., Tirole, J. (1991). "The Politics of Government Decision-Making: A Theory of Regulatory Capture", *Quarterly Journal of Economics*, 106, pp. 1089–1127.

Laffont, J.J. (1994). "Regulation of Pollution with Asymmetric Information", in C. Dosi and T. Tomasi (Eds), *Nonpoint Source Pollution Regulation: Issues and Analysis*, Kluwer Academic Publishers, Dordrecht (Chapter 2).

Mäler, K.G. (1974). *Environmental Economics: A Theoretical Inquiry*, Johns Hopkins Press for Resources for the Future, Baltimore.

Miyao, T., Shapiro, P., Knapp, D. (1980). "On the Existence, Uniqueness, and Stability of Spatial Equilibrium in an Open City with Externalities", *Journal of Urban Economics*, 8, pp. 139–149.

Moretto, M., Graham-Tomasi, T. (1992). "Control of Non-point Source Pollution in a Spatial Setting: A Simplified Approach", *mimeo*, Dipartimento di Scienze Economiche, Universita di Padova, Italy.

Miceli, T.J., Segerson, K. (1991). "Joint Liability in Torts: Marginal and Infra-marginal Efficiency", *International Review of Law and Economics*, 11, pp. 235–249.

Muth, R.F. (1961). "Economic Change and Rural-Urban Land Conversions", *Econometrica*, XXIX (Jan.), pp. 1–23.

Segerson, K. (1988). "Uncertainty and Incentives for Nonpoint Pollution Control", *Journal of Environmental Economics and Management*, 15, pp. 87–98.

Solow, R.M., Vickrey, W.S. (1971). "Land Use in a Long Narrow City", *Journal of Economic Theory*, III (Dec.), pp. 430–447.

Spulber, D. (1985). "Effluent Regulation and Long Run Optimality", *Journal of Environmental Economies and Management*, XII, pp. 103–116.

Tietenberg, T.H. (1974). "Derived Decision Rules for Pollution Control in a General Equilbrium Space Economy", *Journal of Environmental Economics and Management*, 1, pp. 3–16.

Tomasi, T., Segerson, K., Braden, J. (1994). "Issues in The Design of Incentive Schemes for Nonpoint Source Pollution Control", in C. Dosi and T. Tomasi (Eds.), *Nonpoint Source Pollution Regulation: Issues and Analysis*, Kluwer Academic Publishers,

Dordrecht (Chapter 1).

Tomiyama, K., Rossana, R.J. (1988). "Two-Stage Optimal Control Problems With an Explicit Switch Point Dependence: Optimality Criteria and an Example of Delivery Lags and Investment", *Journal of Economic Dynamics and Control*, 13, pp. 319–337.